THE COMPLETE BOOK OF
Insulating

THE COMPLETE BOOK OF
Insulating

Editor Larry Gay

Authors Roger Albright
Larry Gay
Jim Stiles
Eugenia C. Worman
Nathaniel P. Worman
Dana Zak

Drawings by Roger Albright

THE STEPHEN GREENE PRESS
Brattleboro, Vermont

Text copyright © 1980 by Larry Gay.
Illustrations copyright © 1980 by Larry Gay.

All rights reserved. No part of this book may be reproduced without written permission from the publisher, except by a reviewer who may quote brief passages or reproduce illustrations in a review; nor may any part of this book be reproduced, stored in a retrieval system, or transmitted in any form or by any means electronic, mechanical, photocopying, recording, or other, without written permission from the publisher.

This book has been produced in the United States of America. It is designed by Irving Perkins Associates, and published by The Stephen Greene Press, Brattleboro, Vermont 05301.

LIBRARY OF CONGRESS CATALOGING IN PUBLICATION DATA
Main entry under title:

The Complete book of insulating.

    Includes index.
    1. Dwellings—Insulation. I. Gay, Larry,
1937–   II. Albright, Roger, 1922–
TH1715.C615   693.8'32   79-23945
ISBN 0-8289-0364-6
ISBN 0-8289-0365-4 pbk.

Contents

Publisher's Foreword: The Book and the Authors   ix

I   **Making Your House into an Island**   1
Nathaniel P. Worman

    Huff and Puff, 2
    Comfort Is Costly, 3
    Thinking About Insulating, 4
    What Is Heat?, 5
       Temperature.
    Heat Transmission, 5
       Conduction. Convection. Radiation.
    What Is Insulation?, 7
    How Insulation Is Made, 9
    The Economy of Insulation, 10
    Detecting Heat Loss, 10
       Scanning for Heat Loss.
    Insulation as a New Technology, 14

II   **Understanding Heat Flow in Buildings**   16
Jim Stiles

    Degree-Days, 16
    The Thermal Shell, 19
       Air Leaks through the Thermal Shell. Plugging the Holes: Caulking and Weatherstripping. Plugging the Holes: Vents, Chimneys and Outlets. Plugging the Holes: How Far Should You Go?
    Your Thermal Shell and Its R-Value, 23
    R-Values of Various Materials, 23
    Calculating Heat Flows, 28

What Heating Costs, 30
  Cold Bridges.
How Much Should You Insulate?, 32
Insulating Cellar Walls, 35
Solar Energy and Insulation, 36
Heat Transmission through Windows, 37
  The R-Value of Windows. Thermal Curtains and Shutters.
Water and Insulation, 46
  Water Vapor and Vapor Barriers. Precipitation and Insulation. Excessive Humidity.
Fire Safety and Insulation, 51
R-Values of Different Wall Constructions, 52
Notes, 56

III **Weatherstripping and Caulking** 57
Roger Albright

Caulking Windows, 58
Caulking Doors, 61
Caulking Foundation Sills, 62
Caulking the Corners, 63
Caulking the Miscellanies, 64
Caulking Storm Doors and Storm Windows, 65
Masonry Pointing, 67
Kinds of Weatherstripping, 70
Weatherstripping Entry Doors, 74
Checking the Windows, 78
In Sum, 83

IV **Properties of Insulating Materials** 84
Larry Gay

Fiberglass, 85
Other Mineral Fibers, 88
Cellulose, 88
Polystyrene, 89
Polyurethane, 91
Urea Formaldehyde, 93
Reflective Insulation, 96
Vermiculite and Perlite, 97
Notes, 98

v   **Retrofitting Insulation**   99
    Dana Zak

   Materials Handling, 99
   Attic Spaces, 102
      Precautions in the Inspection. Things to Look For. Where to Install Insulation.
   Insulating Unfinished Attics with Blanket Insulation, 105
   Hand-Pouring Loose-Fill in Attics, 108
   Blowing in Loose-Fill Insulation in Attics, 110
   Insulating Finished Attics That Have Knee Walls, 112
   Cathedral Ceilings, Flat Roofs, 115
   Insulating Existing Walls, 116
   Insulating Floors Above Unheated Areas, 117
   Insulating Heated Crawlspaces and Basements, 119
      Internal Insulation of Crawlspaces. Internal Insulation of Basements.
   Hiring a Contractor, 124
      Choosing a Contractor. The Contract. Checking the Work.
   Note, 127

VI **Installing Insulation in New Buildings**   128
    Dana Zak

   External Insulation of the Foundation, 128
      Perimeter and Below-Slab Insulation.
   Insulating Over Unheated Crawlspaces and Basements, 131
   Sill Sealer, 132
   Insulating Walls, 132
      Insulating Cavities in Stud Walls. Installing a Separate Vapor Barrier.
   Insulating Post and Beam Houses, 134
   Fiberglass Inside, Foam Sheathing Outside, 135
   Insulating Unfinished Attics, 136
   Cathedral Ceilings, 137
   Notes, 137

VII **Coping with Other Energy Losers** 138
Larry Gay

    Insulating Hot Water Tanks, 138
        Economical Water Heaters. Standby Losses from Hot Water Tanks. The Economics of Insulating. Installation Details for Electric Heaters. Installation Details for Gas Heaters. Insulating Materials Available. Industry Efforts for the National Cause.
    Pipe Insulation, 149
    Insulating Hot Air Ducts, 151
    Insulating Refrigerators and Freezers, 153
    Notes, 154

VIII **Laws, Government Programs, and Codes Affecting Insulation** 156
Eugenia Worman

    1. Weatherization, 158
    2. Residential Conservation Service, 159
    3. Rural Housing Weatherization Program, 160
    Other Loans and Grants, 161
        Title I Home Improvement Loans. Section 203 (b) and (k) Home Mortgage Insurance. Section 312 Property Rehabilitation Loan. Urban Homesteading. Community Development Block Grant Program.
    Energy Tax Credits, 163
        Credits for Insulation. Credits for Renewable Energy Equipment.
    Rules for Labeling Insulation, 165
    State Plans for Conservation, 166
        Rhode Island. Oregon.

Notes on Further Reading 171

Index 172

PUBLISHER'S FOREWORD

The Book and the Authors

Insulating is an idea whose time has come in no uncertain terms. Whatever your views on our current energy impasse, it is hard to resist the arithmetic demonstrating that the most economical way to make energy available is to conserve it. Insulating makes significant energy conservation possible at relatively modest cost for most householders. Furthermore it does so with no attendant sacrifice of comfort or convenience; on the contrary, the better insulated your living space is, the more comfortable you will be.

As insulating becomes more important, to government at all levels, to industry, and to individuals, new materials, techniques and ideas on the subject multiply. Nevertheless, as the brief Note on Further Reading at the end of this book suggests, the literature of insulating is scanty. In particular not much of the considerable volume of new information on insulants, safety and installation particulars has been published in a form convenient for the general reader.

It is this reader, who wants to be informed on the possibilities for and uses of insulating today—short of becoming a specialist in the field—that THE COMPLETE BOOK OF INSULATING is intended to help. This reader may be a carpenter, builder or other contractor; he or she may be an architect, designer or manager; or the reader may simply be All of the Above—a homeowner or renter who has watched his or her heating and air conditioning bills triple over the last four years.

THE AUTHORS

THE COMPLETE BOOK OF INSULATING assembles a panel of six authors knowledgeable, through practical experience or through formal training—or both—in all aspects of insulating that are important to the reader. All the authors are New Englanders. Some people, including the publishers of this book, believe that the best information on insulating comes from New Englanders, on the same principle that the best information on foxes comes from chickens. New Englanders are closer to the cold and farther from the heat (in the form of oil, gas and coal) than denizens of any other section of the United States. Therefore they stand to benefit more from insulation, and to suffer more from ignorance of it. They know whereof they speak.

Nathaniel Worman. Nat Worman lives in northern Vermont. He edits and produces newsletters for two rural electric cooperative utilities in his region. Through his work for these small coops, both of which stress energy conservation by members, Nat has been involved for some years in issues of the theory and economics of insulating and home heating.

Jim Stiles. Formerly a college student in Vermont, Jim is at present a graduate student at the Massachusetts Institute of Technology. His specialty at M.I.T. is alternative energy systems.

Roger Albright. Roger is an author, editor and artist. He wrote the widely distributed 547 WAYS TO SAVE ENERGY IN YOUR HOME. He has much practical experience of insulating, weatherproofing and building, won in the course of working on a variety of houses, including his own home in Starksboro, Vermont.

Larry Gay. Founder of the L. W. Gay Stoveworks in Brattleboro, Vermont, makers of wood heating equipment, Larry is the author of THE COMPLETE BOOK OF HEATING WITH WOOD, and of a work in progress on central heating with wood and other solid fuels, to be published by The Stephen Greene Press. He has a doctor's degree in chemical physics, and has been a college teacher of chemistry.

Dana Zak. Dana lives in Brattleboro, Vermont, where he has worked on insulation contracting jobs and in the field of alternative energy systems manufacture.

Eugenia C. Worman. Eugenia lives in northern Vermont with her husband, Nat Worman. For the chapter on laws, programs and codes affecting insulation, she organized and summarized information collected from all fifty states and many different offices of the federal government.

THE COMPLETE BOOK OF
Insulating

CHAPTER I

Making Your House into an Island

Nathaniel P. Worman

"When I finish stuffing my house with insulation," said an enthusiastic friend the other day, "I'll light a match at one end of a room and my wife will feel the heat at the other." He said it was the only investment he knew that would yield a 30- to 40-percent return annually—by way of reduced fuel bills. Holding his palms 12 inches apart, he said he was going to put that much insulation in his walls so that they would have an R-value of 36 (referring to the standard measure of a material's ability to block heat flow). And his ceilings, he said, would be *this* thick, with R-50.

Not too long ago, R-11 was considered enough for walls and R-19 for ceilings. In those days our friend would have been called a nut; fuel was so cheap it cost the homeowner less to waste it than to save it. Today there is applause all around for our friend's eagerness. Yet his single-minded interest in insulation could end in bitter disappointment. Having stuffed his walls with insulation, he could still be paying sizable heating or cooling bills, and cursing all the propaganda about insulation he had filled his head with.

For there is far more to insulating than insulation. If you scan the entire structure in which you live or work, you will find an astounding number of vents, holes, cracks, knife-thin slits, and pipes through which heat flows in during the summer and out during the winter. This passage of air in and out—and heat along with it—is called *infiltration*. The significance of infiltration is clear when you consider that the average house has 95 square inches of vents and flues alone; plus window frames, doors, sills, and corners that need sealing and plugging; fireplaces with dampers that never fit; and a front door that is slammed 3000 times a year.

Air passes out through all these openings, as well as in. Therefore *infiltration* is a misleading word—since it denotes a one-way movement of air into a house. We need something better, too, than *breathing* or *respiration* or *exchange*. Perhaps the best way to point to all this flow of air in and out through unsealed slits and unplugged holes is to speak of the *huff and puff* of a house.

HUFF AND PUFF

It is important to realize that by focusing on insulation thickness alone, you are entirely overlooking heat lost by huff and puff. Although you may effectively block the restless flight of heat through floors, walls, and ceilings, you may still, in effect, leave a hole in the wall the size of your fist, for that's what omitting weatherstripping around your front door amounts to. Conversely, an outright and immediate attack on huff and puff, with a modest investment in caulking and weatherstripping, can lead to a savings of 9 barrels of oil yearly, or $500.

The preoccupation with insulation has also blinded us to the devastating effect of windows on our attempts to conserve—not the slit between the sash and the frame or the frame and the house—but the glass panes themselves. A 4 × 4-foot double glazed window will easily lose 18 to 20 gallons of oil a winter, and some studies show windows account for even more heat loss than ceilings.

Double or triple panes are not the only solution. The most effective defense is solid, thermal shutters, put up from the inside, that fit the window hole exactly. They are, of course, abhorrent to many homeowners. Compromises are in order; the shutters can be put up at night behind draperies; they can be kept permanently closed in unused rooms; or, with draperies drawn across them, they can be installed for the entire winter across a glass door that leads to a summer-use patio.

COMFORT IS COSTLY

It's now abundantly clear that the wife of the man with the match in his hand would never feel the heat even two feet away, let alone the whole length of the room, since insulation—stuffing—was his exclusive concern. But tucked away in the man's enthusiasm for insulation is more than his neglect of huff and puff and windows. He is also making a statement about his idea of comfort, and comfort is costly. At the other extreme is the person who insulates with sweater, down jacket and thick socks for life in a house, mainly uninsulated, which is cold or warm in accord with the temperatures outside. Halfway between are most of us, who feel comfortable when the inside winter temperature is at least 65 degrees Fahrenheit (65°F.) and the inside summer reading is about 70. This range is, at any rate, the cultural norm on which this book is based. The cost of comfort in North America is high. Climate plays its part, but so do our great expectations. A look at the history of comfort, and at the word *insulation*, will show how our expectations have risen over the years.

Insulate is from the Latin *insulatus*, which meant "made into an island." The *Oxford English Dictionary* says that the first use of the English infinitive, *insulate*, meant "to make into an island by surrounding with water." This straight-from-Latin sense of the word is the one we like. To insulate your house is to make it an island in a sea of discomforting weather; to insulate it is to reduce huff and puff as well as to put stuffing into the walls.

By 1755 *insulate* meant to cut off or isolate from conducting bodies by the imposition of nonconductors to prevent the passage of electricity or heat. It was not until 1870, in a technical pamphlet, that the word *insulation* was used to denote a material.

The concept of comfort in the modern sense had emerged five hundred years earlier, when more and more homes were built with fireplaces, flues and chimneys to replace the smoky braziers that previous generations had endured. Cold rooms when it was cold outside were a fact of life; it was taken for granted that you would be cold if you walked away from the fire. Why heat the whole structure, after all? There was no indoor plumbing, no threat of frozen pipes bursting.

A uniformly comfortable temperature throughout the whole house was beyond imagination, summer or winter. Central heating and air cooling were unheard of. Decentralized heating meant centralized family life: old pictures of the rural Northeast showing a crowd around the kitchen table meant unheated bedrooms above stairs. In sultry southern Ohio at the turn of the century, wet sheets covered the walls and windows of sick rooms to cool the air. Black electric fans, with blurry faces oscillating, sat on the living room table next to the family Bible. Hand fans flicked so fast you could barely see them. Dawn, with traces of night air lingering, was the time for housework; the severity of indoor temperatures later in the day probably gave rise to the tale of the lady who did all her work just before sunrise, naked.

It is by degrees that our expectations of comfort have risen through the years. If recently we have cut our fuel bills in half, yet continue to heat that unused room, is it fair to accuse us of waste? We seem to be answering that question, in North America at least, by putting a stamp of approval on the idea of a uniformly comfortable temperature throughout a whole structure.

THINKING ABOUT INSULATING

The field of insulating today abounds with inconsistencies, unavoidable where all is so new and information is so incomplete and partly contradictory. Government standards, for example, call for R-19 in all walls in the Northeast. But if the point of insulation is to reduce energy consumption, why not put *less* insulation in the south wall in order to allow it to act as a solar collector when the sun is out? That's a small point, but it underlines the need for a step-by-step logical approach to a field in which it's hard to get your bearings. Here's another point to think about—not a small one. If you insulate your roof and thus keep the snow from melting off, will the snow load bring your building down? *The Complete Book of Insulating* is designed to help you think things out for yourself.

WHAT IS HEAT?

Summer or winter, keeping heat where we want it is the challenge. What is heat? Is it the air around us? We often say we see heat waves. Do we, in fact? We speak of warm breezes or a hot wind. A kettle boils and spews steam and we think Heat. We think Heat, too, when we look at the sun, or touch a piece of metal the other end of which is stuck into a flame. We understand that heat rises: hot air balloons do because the air inside them is lighter than the air outside. Everything about heat seems to be familiar, obvious to the senses, and easily understood.

But the fact is, the nature and behavior of heat are elusive. For our purposes, we can look at three aspects of heat. Each has to do with its movement, each bears on appropriate insulating action, and each is better understood if we see the difference between heat and temperature.

Temperature. Temperature is degree of hotness or coldness on a definite scale. Although we are not used to thinking of temperature in this way, all air around us registers a certain temperature—even if it is 50 below zero—and that means that there is always some degree of heat in the air, unless you can imagine a condition of absolute zero. The gaseous sphere in which we live registers various temperatures at different places on earth and in the sky. If you understand that there can be differences in temperature, you are on your way to getting some understanding of heat. For just as water runs down hill, heat will always run down a scale of temperature differences; flowing from a warmer region to a cooler. Let's look at the three ways it does this.

HEAT TRANSMISSION

Imagine you are sitting comfortably reading this book by your stove on a cold winter day. Your cheeks glow because a moment ago you leaned down, opened the door of your woodstove, and put on another piece of wood. The heat of the flame, while you were doing this, struck your face. And even before that, you had stood on a chair to replace a book on a top shelf and felt how nice and warm it was up there compared to the regions lower down.

Broadly speaking, we all have experienced two things about heat. One is that it travels through a medium (the metal of the stove) and it travels through space (the flame registering heat on your face). Where the medium is concerned, it is more nearly accurate to say that heat energy uses the atoms of the medium for its transmission. For example, when you stick a poker into a flame, the atoms *in the poker* begin to move faster and faster until, dancing madly, they start bumping and accelerating neighboring atoms, which in turn bump others. At last, all the atoms in the poker are moving so fast you drop the poker in alarm. You have burned your hand. Where the heat that you feel on your cheeks is concerned, nothing like that has taken place. The heat from the flame reaches you the same way heat from the sun does, by rays shooting through space without the use of any medium. Radiant energy from the flame reaches you without interacting with air molecules at all. It passes between molecules and even through them.

Here, now, are the three types of heat transfer that are affecting the comfort level in your house as you sit reading this book.

Conduction. This is the migration of heat through matter from atom to atom in, for example, the metal of your stove. The speeding atoms of the flame have jostled and agitated the atoms of which the metal is made. These atoms, in turn, bump each other furiously, thereby transmitting the energy (heat) to the outside surface of the stove. Once there, what becomes of that energy? It is carried off, partly by convection and partly by radiation.

Convection. This is the transmission of heat by movement of matter, in this case, the air just above the surface of your stove. Lighter than the cold air around it, this heated air now rises, carrying its own dancing atoms of heat away from the surface of the stove, thus bringing more energy to a distant corner of the room.

Radiation. Radiation is the transmission of heat, without benefit of a medium, through empty space, that is, in our atmosphere, the space between atoms and molecules. We call it radiant energy because it manifests itself in rays, the source of which are the atoms of which all matter is made. Dancing atoms in your stove fire off radiant energy across the room to excite the atoms of which the wall is made. These atoms now speed up, jostle and bump each other and, by conduction, send the energy on through the wall. It is mainly radiation that heats up the wall next to your stove, which is the reason a shiny material is often used to protect the wall by reflecting the rays.

Convection, conduction and radiation should be the 3-D lenses through which you view the site you select for a new house, and with which you scrutinize the details of correct orientation. For example, what direction does the valley run in which you are building? Do you live where the topography robs you of a significant supply of sunlight during the winter? This will affect the influence of radiation on your house's comfort level. Radiant energy from the sun will also be lost if windows face away from the south, or if the color of your house isn't suitable. There is some evidence that dark houses are more efficient collectors of sunlight than those painted traditional white, primarily because they absorb more radiant energy in the south wall. Considering convection, from what direction do prevailing winds blow, and what natural vegetation can you use to your advantage? Wind (a major convection current) speeds the loss of heat from the outside surface of your house, thereby lowering the temperature on the exterior wall and causing increased conduction through the wall from inside.

Insulating is thus the art of making an island of comfort by managing the conduction, convention and radiation of heat through walls, windows, floors and ceilings.

WHAT IS INSULATION?

In the final analysis, insulation is dead air space, or a dead gas space, sometimes combined with a reflective surface. Air has a low inherent conductivity. If it is dead, or motionless, there is no convection, and when there is a reflective surface, radiation is cut to a minimum. Dead air space can be found in the fluffed-up down of birds in winter. It can be inside a warm jacket or sleeping bag, between layers of blankets, or in insulating materials like fiberglass blanket, loose-fill and foam. All embody thousands of tiny pockets where dead air is encased, protected and preserved.

How effective insulating material is in resisting heat flow is expressed in its *R-value*. This is established in laboratory tests which measure the heat flow through a material in BTUs[1] per hour, per square foot of material, per degree Fahrenheit of temperature difference between the heat source side and the "cool" side. The number obtained from these tests yields the R-value, the units of which—in our unfortunate (English) system—are hour-square foot-degrees Fahrenheit per BTU (hr-ft^2-°F/BTU). R-values are now printed on almost all commercial insulation. Note, however, that listed R-values depend on thickness of insulation as well as on the material used. For example, the R-value of 1 inch of fiberglass is 3.2 hr-ft^2-°F/BTU. A 3½-inch blanket of fiberglass is thus R-11. The higher the R-value, the more effective the insulation.

The foregoing makes it sound as though selecting the best insulating material for your home was a pretty straightforward matter: Buy whatever insulant provides the highest R-value for your money. Unfortunately nothing is ever this simple, at least not in the insulation field today, where there are many unknowns. How much is *not* known about insulation is reflected in a recent action taken by the Department of Energy. DOE has initiated a $40 million research program in the whole field of insulation. What, for example, is the long-term performance in humid climates of fire retardants applied to loose-fill cellulose? Are fiberglass particles a health hazard? The study seeks answers to these and hundreds of other questions, ranging from test procedures for determining the R-value of thick layers of low-density insulation to a look at the combustion products of foam insulation.

But to know for sure how well an insulant performs in real houses lived in by real people will take decades. You may be keen about polyurethane, with its super heat-resistant qualities, for example, and not realize that that high R-value might contribute to starting a fire by preventing heat from escaping from faulty electrical wir-

---

[1] BTU is the abbreviation for British thermal unit. A BTU is the amount of heat that will raise the temperature of 1 pound of water by 1 degree Fahrenheit.

ing. Or a neighbor may talk enthusiastically about urea formaldehyde foam, failing to mention (perhaps not knowing) that formaldehyde is a colorless, noxious gas with a sharp odor, used as a preservative by undertakers and for laboratory specimens. Poorly installed, it will make your house smell just the way you'd imagine.

HOW INSULATION IS MADE

Generally speaking, there are many ways to manufacture insulation which then can be installed as loose-fill, batts or blankets, rigid board, or foam. There is fibrous glass made by subjecting molten glass to blasts of air which blows the material into thin threads. Rock wool is manufactured by melting steel, copper, or lead slag and then spinning it. Vermiculite is made from a mica-like mineral consisting of aluminum-iron-magnesium silicates which, heated to high temperatures, expand and provide a wide range of densities. Perlite is made from natural glass, siliceous and volcanic in origin, the crushed ore particles being expanded from four to twenty times their original volume by rapid heating. Cellulose is manufactured from newsprint, paperboard stock, or virgin wood fiber and is treated with a fire retardant which can wash off if the material is soaked.

Since there is a certain magical appeal to tiny air cells neatly packed away in a virgin white plastic product, it is wise to look at the "poly" insulating materials separately. Polystyrene foam is made into sheets in two ways: by extrusion and by mold, the extruded product having a higher R-value than the molded. Extruded polystyrene (called Styrofoam by the Dow Chemical Company) is made by forcing a hot mixture of polystyrene, a solvent, and a gas through a slit into the atmosphere where, because pressure is abruptly reduced, the mixture expands into a foam with fine, tiny closed cells. In making molded sheets, polystyrene beads are placed in a mold and exposed to heat. This puffs the beads up and fuses them into a large block. Blocks and extruded foam are later cut into sheets for use.

Polyurethane and polyisocyanurate foams are made into solid sheets in much the same ways as polystyrene; or they can be foamed in place. All of the polys are flammable to some degree.

THE ECONOMY OF INSULATION

Since hundreds of laws and regulations have been passed to spur energy conservation in houses, and tax credits and grant incentives are offered, it is wise to ask how much insulation the nation can afford to manufacture, and how much the homeowner can afford to buy. One study suggests that the answer lies in computing the amount of energy it takes to manufacture both the insulation and the walls of which it is a part, plus the amount of energy it takes to operate the house after it is built. By using this kind of approach the study concluded that by the expenditure of an equivalent of 2.1 million barrels of Number 2 fuel oil to manufacture and install insulation, 1.1 million barrels could be saved annually, making the payback in two winters.

Another study offers conflicting testimony. It claims that savings by insulation are grossly exaggerated because the full impact of huff and puff is seldom taken into account: fully two-thirds of heat may be lost through cracks and slits around windows and doors, and up fireplace flues and chimneys.

Once again we are face to face with our ignorance. As a study on insulation by the Department of Energy suggests, we simply do not know enough to make firm decisions about many aspects of insulating, even recommended thicknesses. As fuel prices go up and up, the payback time on investments in insulation will become shorter and shorter. The first 3 inches of insulation, for example, will save a substantial amount of fuel yearly (if a good job of weatherstripping has been done). But with each 3 inches added thereafter savings decline dramatically. Nevertheless, there is clear enough evidence already that, at today's prices of fuel and of insulation, R-19 in the wall is a minimum in northern states.

DETECTING HEAT LOSS

As the discussion of huff and puff above suggested, the amount of heat wasted even by houses well stuffed with insulation would probably surprise their owners. The extent of heat loss from houses is brought home to different homeowners in different ways. One friend told this story. One wintry morning several years ago, a newcomer to the country walked to his kitchen door in answer to furious rattling. When he opened his door to the bright sunlight, several of a flock of chickens huddling there tumbled inside, squawking and pecking. Returning them to their roosts, he found the henhouse door wide open: he had forgotten to slide the spike through the hasp. Later as he chewed breakfast he mulled over a knotty question—why had those chickens backed up against his kitchen door and nowhere else along the sunny porch? Since chickens spend their lives dead from the neck up, it's not likely they would know that a peck or two would make a door open. Maybe, he reasoned, they chose the spot along the back porch that was warmest.

After eating he went to work with a thermometer to prove it. The chickens had huddled there, it turned out, because it *was* warmer outside the door than any other place along the porch, and it was warmer because heat from his oil side-arm inside was flowing out through the door. By noon he had made the outside of his kitchen door tight—weatherstripping was tacked all around, the place where the frame and clapboard met had been caulked, a crack in the door itself was sealed up and a storm door installed. Chickens seeking comfort on a cold morning had spurred to action a man who said that up to that moment everything he had read about insulation was just so many words. Like most of us, he would plug a leak in his gasoline tank the second he saw gas dripping. But energy ghosting through or beneath a door was another matter entirely. The chickens had made him *see* it. So would an infrared scan.

Scanning for Heat Loss. There are two devices that read and record heat loss—the temperature gun and the infrared viewer. You can hire contractors to survey your house for heat loss with either of these pieces of hardware. Choose the viewer over the gun, for two reasons. The viewer will give you an image of the general area aimed at (the whole side of a house, for example) along with the slight temperature differences in varying shades of red (warm) and black (cool). The temperature gun, on the other hand, records by means of a needle only the spot at which it is aimed, and its operator may or may not pick the spot where heat is escaping. In addition, the person with the scanner can, by mounting a camera immediately behind the screen, snap a photo like Figure 1 of your structure.

If you are choosing between two firms who use the scanner, select the one that will provide you in advance with a list of the complete service to be performed: does it include outside as well as inside scanning? The attic? Hot water heater and fuse box (the point of this is to detect electrical shorts)? Will you get a written report? And, finally, don't hire any firm to scan your house which is also selling insulation or caulking or weatherstripping.

Interpreting the infrared scan photo in Figure 1, the white areas show heat escaping from the house, especially through the foundation. The windows are also big losers, even though storm windows are in place. Less heat penetrates the walls, which happen to be plain, uninsulated 2×4 stud walls, typical of houses of this vintage. What you see in the photo is just about what you'd expect if you had already read Chapter 2, where you will learn how to calculate R-values, which in this particular case are about 1, 2 and 3.5 for foundation, double-pane windows and uninsulated stud walls, respectively. Note that heat loss is worse through the narrow windows on either side of the door than through the other windows. The R-value of these single-pane windows is about 1, as can be confirmed by comparison with the greater brightness of the foundation, which indicates a huge heat loss and explains why the crocuses first appear next to the foundation in the spring.

You can buy infrared film for your 35-millimeter camera which will register heat differences. Infrared film, however, is no shortcut or substitute for a proper heat-loss scan like the one in the photo. The film simply does not register temperature differences subtly enough to show heat escaping.

Short of paying for a heat scan, however, there is plenty you can do right now to detect heat loss from your house—even if you don't have chickens. Carrying a rod of lit incense on a mildly windy day, and a pad and pencil, follow your children around inside the house. They'll like the game of finding the places where the smoke bends to reveal air going out or coming in. A wet finger by a window frame tells a similar story, and so does a frozen pipe in your kitchen. Re-

Making Your House into an Island

Figure 1 Pictured below is an infrared scan for heat loss from the house above. The scan shows heat loss to be greater through the house foundation than through the walls or the double-pane windows. (Courtesy of Infra-Scan of Vermont, Shelburne, Vermont)

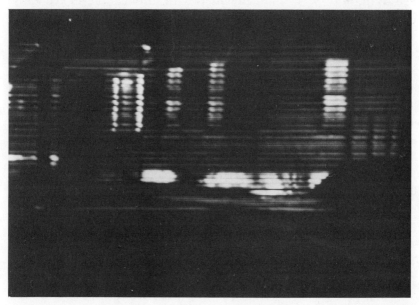

cently in the Northeast, a long cold spell afforded proof of the many botched insulating jobs that exist: there were hundreds of reports of frozen pipes.

INSULATION AS A NEW TECHNOLOGY

Why are we so slow to understand? The enthusiasm for sealing and plugging, whipped up by the charms and benefits of insulation, is relatively new. In fact, the whole field of insulating—as distinct from building—is new itself. A few short years ago insulation was the wallflower in the dance of new energy technologies. Who cared what was in that wall or ceiling when you could fire off a rocket, place a station in space, and make electricity up there without polluting the environment—at least *your* environment. And where in all this was the lab with a million-dollar grant studying anything so mundane as insulation? But now we have elevated insulating to national policy and international significance. The connection with the balance of payments is direct: experts predict that X percent increase in insulating in this country will lead to a Y percent decrease in oil imports. Now freighted with implications of economy and comfort, and with ferociously held dogmas, insulating has become *the* silent and often overlooked weapon in the energy conservation arsenal. Perhaps we are now in danger of overdoing it.

There are forebodings. The British are shying away from humidifiers because they believe tighter houses will rot from too much moisture. Old-time New Englanders would agree. One reason their houses are steady, strong, and old is that unsealed cracks and breaks in the walls allow air to circulate, thereby preserving floor beams, studs, and rafters. Many have wondered how comfortable, both psychologically and physically, a sealed-up house will be. And how healthy? Will extra insulation increase the risk of fire?

How serious is the threat of rot due to moisture in tight houses? Consider the problem. There are showers and baths, washing and drying laundry, cooking and washing dishes, perspiration and breathing, watered plants and humidifiers. A family of four will produce 20 gallons of water in the form of vapor each week. Since water vapor moves through permeable surfaces—floor, wall, and ceiling—a house will breathe out most of those 20 gallons. What re-

Making Your House into an Island

mains, however, is a menace. For as vapor penetrates the cavity of a wall, for example, the dew point is reached and the vapor is deposited as a liquid. Ice can form, weighting down insulation and decreasing its R-value. Later, the wooden frame can begin to rot because of absorbed moisture—a content of only 30 percent moisture will cause wood to rot over a period of years.

What's to be done? Vapor barriers, of course. But they represent only a part of the solution. The other is ventilation. As convection works against you in the contest to keep heat *in* the house, convection works for you by moving air—and moisture—*out* of your attic. If you have installed vents at roof peak, warmer air will be pushed out by denser cold air pushing in through lower intake ports. If you are building, make sure exhaust vents are at the highest and intake vents at the lowest points of the area to be ventilated.

In the insulation war only patience and persistence will prevail. Under huff and puff, the fireplace is a conspicuous glutton of heat, but cracks, slits and small holes are either hidden or don't seem important. Moreover, tracking them down and filling them in often seems impossible. Or, if you do find one (such as the tiny gap around an outside faucet) its insignificance throws the whole enterprise of weatherstripping and caulking into doubt. Big batts of insulation and a snapping staple gun signify real progress. So does cutting wood. Even splitting kindling has its ring of truth and manliness compared to the soft pull of the caulking gun trigger, and the tedious job of making a solid, full bead around window and door frames, down the side of a chimney or wherever two different materials or parts of your house meet. But, if you don't want to waste that wood or make nonsense of that 6 inches of insulation in the wall, it's worth your while to keep at it. Don't worry that the caulk may conduct heat—it will, an insignificant amount—just be pleased it is blocking those convection currents.

If pulling the trigger of a caulking gun makes you feel that there are better ways to use your time, it's wise to remember that you are well in advance of a nationwide trend toward strict conservation building codes for all new houses, and, ultimately, for old structures as well. Insulating is here to stay.

CHAPTER II

Understanding Heat Flow in Buildings

Jim Stiles

One of the most important elements contributing to a comfortable home environment is a moderate temperature. However, when the outdoor environment is either very hot or very cold, maintaining the moderate environment becomes expensive. When you are aware of the high and constantly rising cost of energy, it hurts every time the furnace or air conditioner kicks on. Fortunately it is possible, and often very easy, to significantly reduce the work your heating or cooling system must perform.

DEGREE-DAYS

Before you get into the mechanics of bringing your energy use under better control, you will need to understand one important term: the degree-day (DD) is a somewhat crude measure of how much heat must be supplied to maintain a comfortable indoor temperature for an entire heating season. The degree-day is figured starting at the *balance temperature* of a house, or the outdoor temperature below which a heater must be turned on to maintain a comfortable temperature indoors. Normal household functions, such as lighting and cooking, are secondary sources of heat for the house. When the windows are closed and ventilation is minimized, these heat sources are enough to heat the house by themselves just as long as it isn't too cold or too windy outside. In an average house with the thermostat set at 70°F., no furnace-generated heat is needed until the outdoor temperature drops below 65, the balance temperature in this case.

Weather services use 65°F. as the standard balance temperature for reporting degree-days (Figures 2 and 3). However, in some well-insulated houses the actual heating balance temperature can be as low as 50°F., or even lower in sunny weather.

If the actual balance temperature of your house is 65°F., and the 24-hour average outside temperature is, for example, only 55, then the total for that day would be 10 heating degree-days. If you add up all the degree-days that accumulate throughout an entire heating season, you have a measure of how cold the year was and how much energy you used for heating.

Since many buildings are now air-conditioned, the Weather Bureau also reports cooling degree-days, which are a measure of the

Figure 2 Normal seasonal heating degree-days for North America at base 65°F. (National Oceanic and Atmospheric Administration, U.S. Department of Commerce, Washington, D.C.)

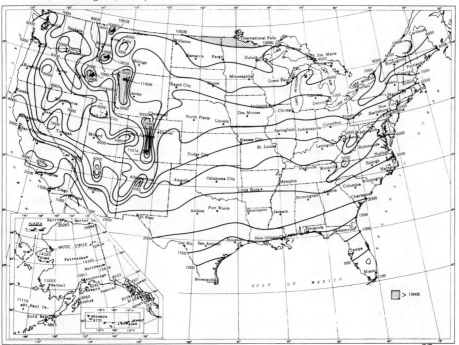

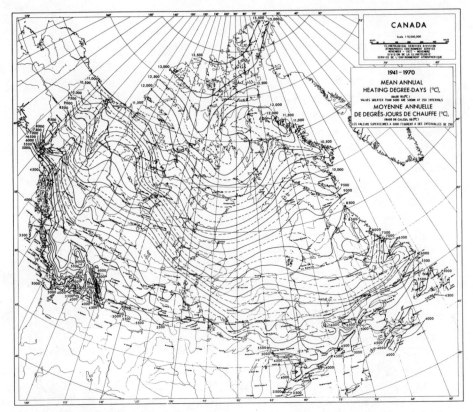

Figure 3 Mean annual heating degree-days at base 65°F. for Canada. (Courtesy of the Atmospheric Environment Service, Canadian Climate Centre, Downsview, Ontario)

energy required in air conditioning (Figure 4). They are calculated by subtracting the cooling balance temperature from the daily average temperature. Here, again, the standard balance temperature is 65°F.

It may seem odd that 65°F. is used as the standard cooling balance as well as the standard heating balance temperature. It *is* odd. After all, most energy-aware people would not turn on the air conditioner until the outdoor temperature reached at least 75 to 80°F. There is in fact a movement afoot among meteorologists to raise the standard cooling balance temperature to 75°F., in keeping with the times.

The number of actual cooling degree-days for two buildings right next to one another can be vastly different, because their actual bal-

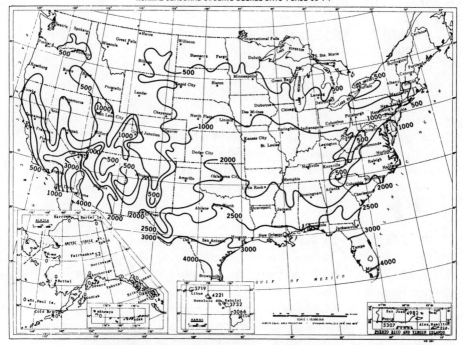

Figure 4 Normal seasonal cooling degree-days for North America at base 65°F. (National Oceanic and Atmospheric Administration, U.S. Department of Commerce, Washington, D.C.)

ance temperatures can be vastly different. If one house has a light-colored roof that reflects sunlight, plus good attic insulation, heat penetration will be much slower. Or if a house is built on a slab and has massive, heat-absorbing walls, these features will keep the house from warming up fast during a hot day. Shade trees and good ventilation also help to raise the actual cooling balance temperature of a house, and these will decrease the number of actual cooling degree-days accumulated during a summer.

THE THERMAL SHELL

A good way to think about your house is to view it as a shell which works to keep heat inside in the winter and keep heat out in the summer. This shell is made up of insulation, interior and exterior sheathing, windows, doors, wooden studs, rafters, bricks, concrete, and other parts that help slow down the flow of heat. The purpose of this shell is to protect you from the extremes of the environment, in much the same way that the shell of an egg protects the life that it houses from its environment. If your thermal shell were perfect, your inside environment would always be at exactly the right temperature, and you wouldn't need a furnace or air conditioner.

There are two ways in which heat passes through your thermal shell. First, heat flows directly through it. (To slow down this kind of heat flow, your thermal shell needs a high R-value, which you get with good insulation.) Second, indoor air, which has been heated or cooled to a comfortable temperature, leaks out through cracks and holes in the thermal shell and is replaced by less comfortable outdoor air.

Air Leaks through the Thermal Shell. The process of air leaking into and out of a house through its thermal shell is called *infiltration* (re-christened "huff and puff" for the purposes of Chapter I). Depending on whether your house is tight and well maintained, or old and drafty (or, heaven forbid, new and drafty), the amount of your energy bill that can be blamed on infiltration can vary tremendously. There is some disagreement among researchers, but generally it is agreed that even in the tightest, best-maintained houses, infiltration accounts for about 20 percent of the heating or cooling bill. Average infiltration losses range between 40 and 65 percent, and of course some houses are even worse. If your house is typical and your heating bill is $800 per year, infiltration is carrying away about 400 of these dollars.

Air leaks are commonly found around windows and doors, between sill plates and the foundation, and other places where there are joints in exterior walls. It is also important to remember that air leaks away through vents and chimneys. Every time your furnace comes on, or you turn on a clothes dryer or bathroom fan, you lose hundreds or thousands of cubic feet of interior air, which is replaced by air from outside. Even when the blowers and burners are not operating, vents and chimneys provide easy routes for air to leak out of a house.

Plugging the Holes: Caulking and Weatherstripping. Weatherstripping is one of the two best energy-saving investments you can make. The other is caulking. Sometimes cracks and joints will be so large that caulking and weatherstripping will not suffice. In these cases some carpentry is called for.

Some cracks and joints will be obvious—you will be able to see them, or feel drafts through them on breezy days. Obviously, all such joints and cracks should be taken care of. Other less obvious leaks will probably slip by undetected.

In order to find these less obvious leaks, proceed as suggested in Chapter I. On a windy day, take a smoldering cigarette, stick of incense, or other source of smoke and go hunting. Use the movement of the smoke to precisely track down the leaks, then plug them with the appropriate weatherstrip or caulk. Chances are that you will find leaks in some pretty peculiar places, so be sure to check around electrical outlets, electrical fixtures, cracks between pipes and walls, and other places you might not expect to find air leaks.

Doors present a special infiltration problem. You can caulk around the door frame, and weatherstrip the door so that it seals well when it is closed. However, especially for people with children and pets, the door spends a lot of time open. Using the house itself to block the wind helps reduce air blowing through an open door, but sometimes the layout of the house dictates a door on the windward side. Trees and shrubs may also help to block the wind, but the ultimate answer for this problem is an enclosed vestibule. Just as long as one door of the vestibule is closed when the other is open, there will be an air lock.

Plugging the Holes: Vents, Chimneys and Outlets. It is common for a bathroom or kitchen fan to exhaust 100 cubic feet of air per minute. In an hour's time this totals 6000 cubic feet, equivalent to exchanging something like one-half of all the air in a house every hour.

Even when fans, blowers and burners are not operating, the simple fact that vents and chimneys are holes in the thermal shell means that air will leak through them. Dampers in fireplace flues are seldom tight, nor do clean-out doors at the bottom of chimneys always fit snugly. Vents can be outfitted with louvers that close automatically, and automatic dampers are now available to be installed in the flue of an oil or gas furnace as a retrofit item.

Another approach to reducing heat loss through vents and chimneys is to make sure that you use them only when you need to, or at the most sensible time. For example, on cold days, run your dryer in the early afternoon when it is relatively warm so that the air that leaks in to replace what is blown out is at least not quite as frigid as it might be. In the summertime, run the dryer in the morning or evening, so that cool morning or evening air is drawn in to replace air the dryer blows out. If you are installing a new vent or replacing an old one, make sure that the fan is no bigger than absolutely necessary. Another idea is to run all air from one or more exhaust vents through an air-to-air heat exchanger. In this system the outgoing air heats, or cools, incoming air.

Electrical outlets may also be a source of infiltration. Cold air coming in through an electrical outlet is often best combatted at the point where the wires enter the wall in the attic or basement, rather than at the outlet itself.

Plugging the Holes: How Far Should You Go? Beyond plugging up all the cracks and joints you can find, it is often difficult to determine whether an improvement of your thermal shell will save you more than it costs. There are two valuable basic perspectives that may help you decide.

Probably most important is that if you are going to do some home improvement anyway, it is an excellent time to do something about infiltration. For example, if you are installing a new furnace, spend a little extra time to make sure that the flue connector is absolutely tight so air from the basement cannot sneak into the chimney. Or, if you have to install new steps to your main door, think about adding a vestibule at the same time. It will probably be the best chance you ever have.

The other perspective is that if you make your house really tight, you may notice that it is not quite as fresh smelling as it was before. Odors from the garbage can, cooking and laundry hamper begin to linger. This is a sure sign that you have had some success in controlling infiltration. In a clean house, such odors mean that the exchange rate is well below one air change per hour.

If odors do become a problem, there are three solutions: (1.) Eliminate or seal off the sources of odors; (2.) Air the house on warm winter days or cool summer nights; or (3.) Install a filtering system that recirculates interior air through a charcoal filter. Green plants also help. In any case, let odors serve as an indication of your progress. When the house becomes unpleasantly stuffy, it may be a sign that your house is tight enough.

YOUR THERMAL SHELL AND ITS R-VALUE

A good way to think of your thermal shell is to consider it as being like the clothes you wear, since one of the main functions of clothing is to act as a thermal shell. Installing thick insulation is like putting on a goose down jacket.

Different insulating materials have different R-values. Fiberglass, cellulose fiber and other fluffy fibrous materials are all similar, with R-values of around 3 for each inch of thickness. Plastic foams like polystyrene, urea formaldehyde, and especially urethane and isocyanurate have higher R-values per inch. Vermiculite and perlite are lower.

R-values of all materials are subject to some variation, depending on density of the material, water content, and even temperature. Thus one runs across reported R-values for fiberglass batts from 2.9 to 3.5 per inch, depending on the source. In fact, there is some inconsistency among R-values in this book. That one reported value is higher than another for the same material does not mean someone is trying to fool you. We know of instances where laboratory R-values are actually higher than the nominal value reported by the manufacturer. On the other hand, R-values in many actual houses may be substantially lower than nominal because of the effect of moisture. This is especially true of retrofit installations where no decent vapor barrier exists.

R-VALUES OF VARIOUS MATERIALS

Most forms of insulation trap many small pockets of air, or—in the case of plastic foams—other gases. Pockets of air slow down the flow of heat partly because still air is intrinsically a poor heat conductor and partly because convection is suppressed in small pockets of air. As we said, fiberglass has an R-value of about 3 per inch. However, the glass from which the fibers are made has an R-value of under 0.2 per inch. Because the glass fibers trap air pockets, the R-value is increased 15 times. Generally speaking, the smaller the pockets of air, the higher the R-value.

Trapping layers of air is not as effective as trapping pockets of air, but sometimes it's the best you can do, as in the case of storm windows or commercial double-pane windows. The R-value of a layer of air increases up to a width of about ¾ inch. Beyond this point, convection currents circulate freely between the two surfaces and carry heat from the hot side to the cold side, as Figure 5 suggests, so that further increases in width no longer improve the R-value of the gap. Air layers that are greater than ¾ inch in thickness but less than about 4 inches all have R-values very close to 1. An air space is slightly more effective in the floor than in the ceiling because natural convection carries heat up in the ceiling but won't carry it down within the floor.

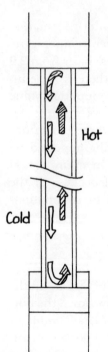

Figure 5 Heat is transferred from the warm to the cool side of a double-pane window's interior by convection of air between the panes.

A point to appreciate is why convection doesn't destroy all the R-value of an air space. The reason is that air has a tendency to adhere to surfaces. Thus, thin layers of air cling to windows, ceilings and floors. As long as there aren't great gusts of wind, the film of air adhering to a surface has an R-value of about 0.7 (strong winds can reduce it below 0.2). This also explains why many small pockets of air insulate a given surface area (a wall, say) better than a few large air spaces insulate an equal area: small pockets mean relatively more total enclosed surface area for air to cling to, and so more of the trapped air is stagnant and less is moving.

One way to significantly improve the R-value of an air space is to use shiny metallic surfaces like aluminum foil. This works well because a large part (typically about one-half) of the energy flowing through an air space is carried by infrared radiation, which passes easily through air or a vacuum. A shiny metallic surface generates very little infrared radiation and also reflects most infrared that strikes it. The reason metallic surfaces aren't used more is that they eventually become dull and lose effectiveness.

A good way to become acquainted with R-values is to take a leisurely look at Table 1.

TABLE 1
R-Values

	R per inch thickness (hr-ft^2-°F/BTU)
Still Air (conduction only)	5.9
Freon 11 (in urethane foam, conduction only)	20.8
Glass Wool	
batts and blankets	3.1
loose-fill	2.3
high-density	4.2
Mineral wools other than fiberglass (rock wool and slag wool)	
batts and blankets	3.4
loose-fill	
1.5 pcf (pounds per cubic foot)	2.1
2.5 pcf	3.4
3.5 pcf	3.9

TABLE 1 (Continued)
R-Values

	R per inch thickness (hr-ft^2-°F/BTU)
Cellulose	
loose-fill	3.7
spray-on	4.0
Beadboard polystyrene	
1 pcf	4.0
1.5 pcf	4.2
2.0 pcf	4.4
Extruded polystyrene	
1.75 pcf	4.4
2–2.4 pcf	5.5
Polyurethane	6.3–7.5
Isocyanurate	7.0–8.0
Urea formaldehyde	4.0–4.3
Asbestos wool	1.4
Vermiculite	2.0
Perlite	
2 pcf	3.7
5 pcf	3.1
10 pcf	2.6
Siding and Sheathing Materials	
wood clapboards or shingles	1.00
stucco	.20
plywood	1.25
fiberboard	2.1–2.8
hardboard	.70
softwood	1.25
hardwood	.90
gypsum board	.90
Structural Materials	
concrete block (figures for *total* R)	
4 inches	.70
8 inches	1.20
12 inches	1.35

TABLE 1 (Continued)
R-Values

	R per inch thickness (hr-ft²-°F/BTU)
concrete	.10
lightweight concrete	.35
steel	.002
aluminum	.001
brick	
common	.15–.20
face	.10–.15
lightweight	.20–.35
stone	.05–.10

Miscellaneous

asbestos millboard	1.20
plaster	.60
corrugated cardboard	2.25
corrugated asbestos paper	2.70
glass	.15
carpet plus fibrous pad (*total* R)	2.10
carpet plus foam pad (*total* R)	1.25
ceramic tile	.15
concrete tile	.15
cork	2.00
built-up roofing	.90

Air Spaces (figures for *total* R of 3 thicknesses)

¾–4 inches (plain surface)	1.0
¾–4 inches (one reflective surface)	2.05
¾–4 inches (two reflective surfaces)	2.25

Air Surfaces, non-reflective (figures for *total* R)

horizontal (heat flow up)	.60
horizontal (heat flow down)	.90
vertical (still air)	.70
vertical (15 mph wind)	.15
vertical (7.5 mph wind)	.25

CALCULATING HEAT FLOWS

If you know what the R-values of the various layers in a thermal shell are, it is an easy matter to calculate the total R-value (R_{tot}). The total R-value is the sum of all the R-values that go into making up the insulating shell.

$$R_{tot} = R_1 + R_2 + R_3 + \cdots$$

Figure 6 shows how to find the total R-value for two common parts of the thermal shell: a fiberglass-insulated exterior wall and a double-pane window. Other practical examples are given at the end of this chapter in Figures 19–24. From the second example in Figure 6, it is obvious that the insulating value of the glass itself is negligible. Even with 2 layers totaling ¼ inch in thickness, the glass counts only for 2 percent of the total insulating value. For this reason, most calculations which involve window glass, aluminum window

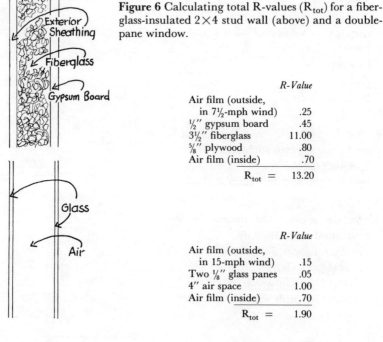

Figure 6 Calculating total R-values (R_{tot}) for a fiberglass-insulated 2×4 stud wall (above) and a double-pane window.

	R-Value
Air film (outside, in 7½-mph wind)	.25
½″ gypsum board	.45
3½″ fiberglass	11.00
⅝″ plywood	.80
Air film (inside)	.70
R_{tot} =	13.20

	R-Value
Air film (outside, in 15-mph wind)	.15
Two ⅛″ glass panes	.05
4″ air space	1.00
Air film (inside)	.70
R_{tot} =	1.90

Understanding Heat Flow in Buildings

frames, or other thin layers of low-R materials simply do not include the contributions of these layers.

Once you have the total R-value of a section of the thermal shell and the temperatures on each side of the section, you can calculate the amount of heat flowing through the section. Just use *The Basic Heat Flow Equation*:

$$q = \frac{A \times (t_1 - t_2)}{R_{tot}}$$

where q = rate of heat flow, BTU/hr.
A = the area of the section, square feet.
t_1 = the warmer temperature, °F.
t_2 = the colder temperature, °F.
and R_{tot} is what you just calculated.

Looking at Example 1 in Figure 6, suppose you want to know the rate at which heat is flowing into the house through *one square foot* of the wall, and suppose the outside temperature is 90°F. and the inside temperature 75°F.

$$q = \frac{1 \times (90 - 75)}{13.2} \text{ BTU/hr}$$

$$= \frac{1 \times 15}{13.2} \text{ BTU/hr}$$

$$= 1.14 \text{ BTU/hr.}$$

Now look again at the second example, where $R_{tot} = 1.9$; assume that the outside temperature is 15°F. and the inside temperature is 70, and that your window is 3 × 5 feet.

$$A = 3 \text{ ft} \times 5 \text{ ft} = 15 \text{ ft}^2$$
$$q = \frac{15 (70 - 15)}{1.9} = \frac{15 \times 55}{1.9} \text{ BTU/hr}$$
$$q = 434 \text{ BTU/hr.}$$

If you want to figure out total heat loss through this double window over the course of a heating season, you must also know the figure for the degree-days in your area. Use the following modified version of the Basic Equation.

$$Q = \frac{DD \times 24 \times A}{R_{tot}}$$

where Q = BTU flowing through the window every year
DD = degree-days (of heating or cooling)
and 24 is the conversion factor for days to hours.

We are now in a position to calculate the heat loss through the double window during an entire year. Suppose DD = 8000. This would be a cold climate, e.g., Madison, Wisconsin, or Burlington, Vermont.

$$Q = \frac{8000 \times 24 \times 15}{1.9} = 1{,}515{,}000 \text{ BTU/yr.}$$

Calculating the heat flow over the course of a year may be in error if you use the number of degree-days reported by the Weather Bureau and the balance temperature of your house is not 65°F. The error can be sizable. For example, with a balance temperature of 65°F., the number of degree-days in Madison is 8000, but with a 60-degree balance temperature, the number falls to 6627, a 17-percent difference. This 17-percent error would be incorporated into the heat flow calculation by using the standard number of degree-days reported by the Weather Bureau.

WHAT HEATING COSTS

In the window example of Figure 6, the heat loss amounts to 1,515,800 BTU/yr for each double-pane window. If the heat source is oil (fuel value = 140,000 BTU/gallon) and the furnace is 60 percent efficient, then 18 gallons' worth of heat go sailing out through each such window each winter. At $1.00 per gallon, that's $18 per window:

Understanding Heat Flow in Buildings

$$\frac{1{,}515{,}800}{0.60 \times 140{,}000} = 18 \text{ gallons}$$

A single-pane window having an R-value of 1 instead of 2, would lose 36 gallons of oil a winter ($36).

Cold Bridges. In an insulated stud wall, the studs themselves have a significantly lower R-value than the wall sections containing insulation. This type of low-R element penetrating a thermal shell is known as a cold or thermal bridge. Its effect can be seen in Figures 7 and 8. The stud section of the wall has a total R-value of 7.00, and the insulated portion an R-value of 13.45. If the framing constitutes 20 percent of the total wall area, the *effective* R-value of the wall is 11.35. (See Figures 8, and 19-24, for the mathematical details.) This is to say that the studding in this wall reduces the overall R-value by 2.1 (16 percent) compared to the totally insulated wall. Note that the R-value of the fiberglass alone is R-11. When someone speaks of an R-11 wall, he usually means that the cavities between studs are filled with R-11 insulation. In practice the effective R-value of the wall taken as a whole is often nearly the same as the R-value of the insulation alone. Therefore, often it is not far wrong

Figure 7 Insulated stud wall seen from above, showing a cold bridge. The wood studs offer much less resistance to heat flow than fiberglass insulation.

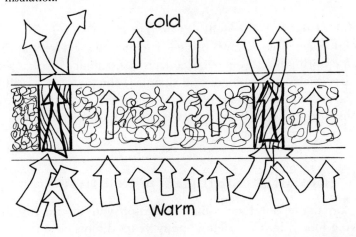

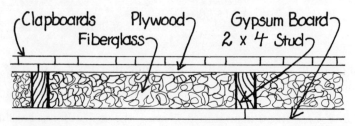

Figure 8 Calculation of total and effective R-values of an insulated 2×4 stud wall.

	R-values in Framed Section	R-values in Insulated Section
Air film (outside, in 15-mph wind)	.15	.15
Clapboard siding	.80	.80
⅜" plywood	.45	.45
R-11 fiberglass blanket	—	11.00
3⅝" stud	4.55	—
⅜" gypsum board	.35	.35
Air film (inside)	.70	.70
R_{tot} =	7.00	13.45

If the framing constitutes 20 percent of the total wall area, then the *effective* R-value,

$$R_{eff} = \frac{1}{\frac{.80}{13.45} + \frac{.20}{7.00}} = 11.35$$

to think of the wall thermally as though it were insulation alone.

Another example of cold bridges is concrete piers beneath rock beds for storing heat from solar collectors. Ordinary insulants cannot carry such a heavy load, and therefore the piers are absolutely necessary. A check of the table of R-values will show that concrete has one of the lowest of all R-values listed, and the piers thus provide a path for the escape of a significant amount of heat.

HOW MUCH SHOULD YOU INSULATE?

You can figure out how much insulation you should install by knowing four things: (1.) How many years the insulation will last

Understanding Heat Flow in Buildings

before it must be replaced; (2.) How heat loss depends on R-value; (3.) The cost of heat; (4.) The cost of the installation.

If you add the cost of the heat that flows through an insulated section and the cost of installing the insulation, you end up with a total cost of keeping your house comfortable. If you repeat this process for many different types of insulation and different thicknesses of insulation, you can compare them and select the lowest cost.

Of course the whole process is filled with uncertainty. How long the insulation will last depends on things like how well you maintain the roof and whether mice move in or not. The cost of heat is unpredictable, except to say that the trend is up. You could, if you wanted, also take into account alternative investments or the cost of borrowing money to pay for materials and labor. However, given all the uncertainty, there is no point in trying to refine the calculation very much. Remember that today's economical R-value is apt to be too low tomorrow.

Now to the calculation of cost-effectiveness. First, it is necessary to know how much you improve the performance of your thermal shell by installing extra thicknesses of insulation. For any given set of conditions, the results will give a curve having the same general shape as Figure 9. This curve shows that each additional inch of insulation, regardless of kind, brings diminishing returns. Since each BTU consumed by your furnace or air conditioner has a particular cost, the curve for cost of energy vs. inches of insulation has the same shape. You simply use dollars instead of BTUs.

To get the total cost of fuel plus insulation, you must also calculate carpentry costs and the cost of the insulation itself for various

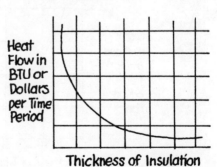

Figure 9 Graph shows that each added inch of insulation is less effective in reducing heat loss and fuel bills than the one that went before.

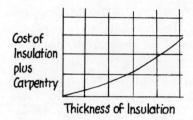

Figure 10 How the cost of insulating depends on thickness of insulation.

thicknesses of insulation. The graph of these costs looks something like Figure 10.

Now, if you add the cost of fuel plus the cost of insulation and carpentry and graph that against inches of insulation, you get a result that looks like Figure 11.

Figure 11 happens to be for stud-wall construction filled with fiberglass and subjected to 6000 degree-days. Its message is that 12 inches of fiberglass (R-36) leads to the lowest total outlay under our assumptions. We have used projected 1980 oil prices in the computation and fairly cheap labor. Had we let the price of oil escalate or had we borrowed money at a high rate, the optimum amount of insulation would of course be changed somewhat. However, note that the curve does not have a sharp minimum. That is, you will generally have a fairly broad range where adding or subtracting a few inches of insulation won't make much difference to the total cost of keeping warm. Eight or 16 inches of fiberglass would be almost as good as 12. Thus the choice is not crucial, and must be

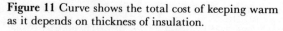

Figure 11 Curve shows the total cost of keeping warm as it depends on thickness of insulation.

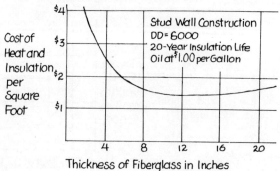

influenced by common sense. In our example one can easily justify only 6 inches of fiberglass, for several reasons: (1.) There is a big cost jump in going from 2×6 studs to larger ones; (2.) If thermostat setback is taken into account, the minimum in Figure 11 is shifted to the left; (3.) The capacity of the country to produce fiberglass is limited.

The calculation for a ceiling instead of a wall proceeds in just the same way. A big difference, however, is that the optimum thickness for attic insulation is larger than for walls—not because "heat rises" but because extra carpentry costs are not incurred by adding extra inches of insulation to a ceiling. True, convection currents can carry hot air up into the attic, but the way to stop these is not with many inches of fiberglass or cellulose. Rather, it should be by plugging air leaks through light fixtures, around chimneys and attic doors, etc.

You can, if you are so inclined and good with figures, make calculations like the foregoing for yourself to help determine the economically best insulation for your house. However, it must be emphasized again that the whole business is fraught with uncertainty because the assumptions are so uncertain. For most people it is much easier to call the county extension agent or the state energy office to see what R-values they recommend for a particular area. Their estimates are based on the same kind of calculation and are apt to be fairly close to the mark.

INSULATING CELLAR WALLS

Only in a very few homes are the cellar walls insulated. But in cold climates they should be. A glance at the table of R-values (Table 1) will disclose that any form of masonry has an abysmal R-value. In warm climates, where cooling is more important than heating, basement walls should generally not be insulated, since insulation of the basement detaches the house thermally from the heat-sink the earth provides.

Methods of insulating subsurface foundation walls are discussed in Chapter VI. We are concerned at this point with quantity of below-ground insulation. Unfortunately, this is a tough decision to make—R-values of different soils vary greatly. A big factor is ground water. Wet soil has a much lower R-value than dry soil. But even more important is the question of whether or not the ground water is flowing along the foundation. If there is no flow, heat escaping from the house warms the soil, decreasing the need for insulation. However, even small flows of ground water quickly carry heat away and make insulation that much more important. The place to start, therefore, is with improving drainage away from the foundation.

Roughly speaking, walls of heated basements in cold climates require about one-half to two-thirds as much insulation as do walls above ground. When the basement is not heated directly (only through furnace losses) the basement walls should have an R-value roughly one-third of that of the walls above ground. Extruded polystyrene is a good material to use. Many people use high-density (2 pounds per cubic foot) beadboard for basements because it is cheaper, but it may not last as long as the extruded product. In a climate of 6000 degree-days, approximately 3 inches of extruded polystyrene is adequate. In climates where heating and cooling loads are similar, about 1 inch of foam is called for on most of the cellar wall, with a little more at the top.

SOLAR ENERGY AND INSULATION

So far we have given the straight, unembellished story on insulating, as if it were a simple matter of plugging in the numbers and turning the crank. Now for something interesting.

As the sun beats on the south side of a house during the daylight hours of the winter, the wall can be warmed significantly. In fact, in winter the south wall, forgetting the windows for the moment, can act as a low-grade solar collector with an inward flow of heat when the sun is out. Recent work has shown that even the infrequent and diffuse sunlight in cloudy Oregon causes more than a one-third reduction in heat loss through a south wall filled with R-19 fiberglass.[1]

Clearly this effect should be taken into account when figuring the amount of insulation for the south wall. Insulating rules that do not single out the south wall for special attention can lead to excessive insulation of the south wall and even reduce its effectiveness as an energy flow regulator. Sunny New Mexico has taken the lead in allowing deviations from standard codes for the south wall.

Even though there is no way that the implications of this solar effect can yet be analyzed precisely, we can still make some general rules. First, when calculating the optimum amount of insulation for the south wall, don't use the standard heating degree-day for your region. Instead, decrease it by some factor. This factor is obviously going to be very crude, unless you happen to live near Portland, Oregon, where it is $2/3$. The multiplier will tend to be lower in sunnier regions, and lower yet for buildings with dark-colored south walls—which are the best for absorbing solar energy.

If the sun's effect on the south wall is to reduce the need for insulation there, then we can expect extra insulation to be required on east and west walls—which absorb a great deal of sunlight in summer when you don't want the heat. The multiplier to use with cooling degree-days will be greater than 1, since absorbed solar energy puts an extra burden on the air conditioner in summer.

In any event, begin thinking of the south wall as different from the rest.

HEAT TRANSMISSION THROUGH WINDOWS

A window on the south side of your house is a medium-grade solar collector. On any other side of the house, you might envision a window as a thermal Black Hole, with a great many very expensive BTUs flowing freely through it. As an element of your thermal shell, a window can be either a tremendous asset, or a terrific liability.

Over the course of a winter, no matter where you live in North America, each square foot of double-glazed window on the south side of your house will collect approximately the amount of heat that one gallon of oil would give you. During the summer a south window passes less than half the energy it does in winter, because the summer sun is high in the sky so that only a small amount of solar energy strikes the south windows. Most of what does hit is reflected, because it strikes at a steep angle. A 2-foot overhang will shade south windows completely in summer. South windows give you the best of both worlds—solar heat all winter and soft pleasant light in summer.

East and west windows collect virtually no solar energy in winter, but they "turn on" and collect energy all summer. North-facing windows never collect a significant amount of energy in the northern hemisphere, except near the North Pole in summer.

The R-Value of Windows. As mentioned above, the air films that adhere to the inside and the outside surfaces of a single-pane window give it an R-value of about 0.9. Strong winds and extreme temperatures, which create strong drafts next to the glass, both tend to lower the R-value. Each additional ¾ inch or larger air space trapped by extra layers of glass adds another R-0.9 to R-1.0. If you live in a very hot or very cold climate, it will pay to install several layers of glass in order to develop a better total R-value. Here, though, we should anticipate a point in the discussion to come, and observe that window shutters will be a much better investment.

Using Figure 12, it is easy to find how much heat you lose through a window. You simply calculate the window's R-value and estimate the number of degree-days of heating in your climate (refer to Figure 2 or 4). Then you select the curve that is the closest match to your number of degree-days, and find where your window's R-value intersects the curve. Finally, read off how many gallons' worth of heat flow through each square foot of window. If

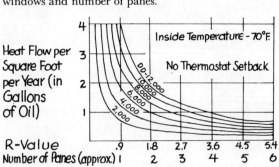

Figure 12 Relationship between heat loss through glass windows and number of panes.

Understanding Heat Flow in Buildings

you don't heat with oil, you can use Table 2 to convert to your fuel. Finally, you can convert your actual gallons (or other energy) used into dollars per square foot of window per year, or better, dollars for the total window over the total lifetime of the window.

For example, in Boston, we have 6000 degree-days of heating. I use oil to heat my home, and I want to figure out what to do with my windows. I start calculations on one single-glazed 3 × 5-foot window. In Table 1 we see that the R-value of a vertical interface between glass and still air on the inside is about .70, and between glass and moving air on the outside (15 mph) is about .15.

$$R_{tot} = .15 + .70 = .85$$
$$DD = 6000$$

Reading from the graph in Figure 12, I find that just over 2 gallons' worth of heat flow through one square foot of that window every year. For all 15 square feet of glass, and over the 30 years that I figure the window will last, it will leak 900 gallons of oil.

TABLE 2
Fuel Conversion Table

	Fuel Value	Assumed Efficiency	Available Heat	Equivalent of 1 Gallon of Oil
oil	140,000 BTU/gal	60%	84,000 BTU/gal	—
natural gas	1000 BTU/cubic ft	65%	650 BTU/cubic ft	129 cubic ft
bottled gas	2500 BTU/cubic ft	65%	1625 BTU/cubic ft	52 cubic ft
electricity (resistance heating)	3410 BTU/kwh	100%	3410 BTU/kwh	25 kwh
electricity (heat pump)	3410 BTU/kwh	200%	6820 BTU/kwh	12 kwh
hardwood	28 million BTU/cord	55%	15.4 million BTU/cord	.0055 cord

Converting from gallons to dollars is not going to be wholly accurate. If I guess that oil will increase at about 3 percent a year faster than inflation, a correction multiplier must be taken into account, in this case 1.58 (See Table 3). The present price is about $1.00 per gallon, so that the total cost of energy lost through the window will be $1.00 × 1.58 × 900 = $1422 in current dollars.

TABLE 3
Multiplier for Taking Real Energy Price Increase into Account

Annual Increase in Cost of Energy %

Years	1	2	3	4	5	6	7	8	9	10
5	1.02	1.04	1.06	1.08	1.10	1.13	1.15	1.17	1.19	1.22
10	1.05	1.09	1.15	1.20	1.26	1.31	1.38	1.44	1.51	1.58
20	1.10	1.21	1.34	1.49	1.65	1.83	2.04	2.27	2.53	2.83
30	1.16	1.35	1.58	1.86	2.20	2.61	3.12	3.73	4.47	5.38
40	1.22	1.51	1.88	2.37	3.00	3.83	4.92	6.37	8.28	10.80

If I add another sheet of glass separated by a ¾-inch air space, the new R-value will be:

$$R_2 = R_1 + 1 = 1.85$$

From Figure 12 I find that I will lose about 0.95 gallons per square foot per year, or 427.5 gallons for the entire window over 30 years. This converts to $675.45, or a savings of $746.55.

I can keep on calculating the savings realized by adding more sheets of glass (and layers of air) until the savings for adding another sheet are less than the cost of adding that sheet.

It might be objected that each layer of glass added to windows reflects some solar energy. This is true, but only important in the case of south windows. Roughly speaking, each pane of glass reflects about 10 percent of the solar energy striking it. The second pane of glass in a south window reduces heat gain by another 10 percent, but at the same time it reduces heat loss by 50 percent. Thus multiple glazing of south windows is justified, too.

You now know how to calculate the cost of the heat, and you can either estimate installation costs yourself, or have a carpenter do it. For most people, installation estimates for two, three and four layers of window glass will suffice, since two layers are justified anywhere in the United States, and almost no one will want more than four layers.

Thermal Curtains and Shutters. One of the big problems with installing many layers of glass in a window is that glass is relatively

Understanding Heat Flow in Buildings

expensive, especially considering the fact that each layer gives you only another R-0.9 to R-1. The way to get more R-value for your dollar is to use thermal curtains or shutters. The big drawback is that they require frequent attention.

A thermal curtain is a curtain that is especially designed and installed to improve the thermal performance of a window. It will usually include a layer of flexible insulation, often in the form of a quilt, and may also have one or more shiny, reflective surfaces.

The "High-R Shade" in Figure 13 is just one example of the many recent innovations in thermal curtains. The multiple reflective layers give it an R-value of at least 9, and probably more. In Figure 14, we see that the R-value for the triple-glazed window covered by a one-inch quilt is 6.60, compared with 2.65 for the uncovered window. If you're really serious about saving energy for the least expense, then use a urethane or isocyanurate shutter, espe-

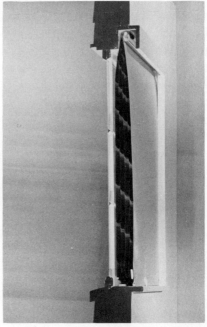

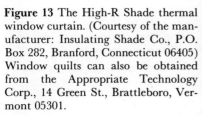

Figure 13 The High-R Shade thermal window curtain. (Courtesy of the manufacturer: Insulating Shade Co., P.O. Box 282, Branford, Connecticut 06405) Window quilts can also be obtained from the Appropriate Technology Corp., 14 Green St., Brattleboro, Vermont 05301.

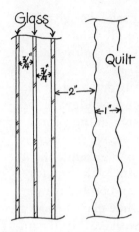

Figure 14 R-value of triple-pane window with and without a quilted thermal curtain.

	R-values Without Curtain	R-values With Curtain
Air film (outside)	.20	.20
Two ¾" air spaces	1.80	1.80
2" air space	—	.95
1" quilted curtain	—	3.00
Air film (inside)	.65	.65
R_{tot} =	2.65	6.60

cially over single panes where you can almost see the oil leaking out. Figure 15 shows that a 1-inch urethane shutter increases the thermal resistance of the single-glazed window by more than 900 percent. Note, however, that the National Fire Protection Association does not approve of the use of urethane or isocyanurate without a fire barrier of ½-inch gypsum board or the equivalent because of toxic fumes released if the foam burns. Gypsum board is too heavy to use on a shutter of this type, and no inexpensive lightweight equivalent has yet been developed, so here—unlike the 55 mph

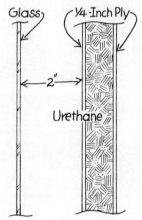

Figure 15 R-value of single-pane window with and without urethane thermal shutter (plywood sheathed).

	Without Shutter	With Shutter
Air film (outside)	.20	.20
2" air space	—	.95
1" urethane	—	6.30
2 layers ¼" plywood	—	.60
Air film (inside)	.65	.65
R_{tot} =	.85	8.70

Understanding Heat Flow in Buildings

speed limit—we have a real and unresolved conflict between safety and energy conservation.

The best thermal curtains are designed so that there is a good seal between the curtain and the window frame. However, contrary to the opinion of many, a good seal at this joint is not critically important for good thermal performance. If you simply make a modest effort to make the joints tight, results will be excellent, *as long as the window itself is tightly weatherstripped and not leaky.* A series of experiments by William Shurcliff has demonstrated clearly that the elaborate sealing systems employed with some curtains can help, but in practice, the insulating effectiveness is nearly the same for an excellent seal and for a moderately sloppy fit.[2] (The best reason to try for a tight fit is to keep water from condensing on the window overnight.) Anyone who is reasonably handy around the house can make window curtains or shutters for himself that will be—next to a tube of caulking compound—about the best investment anyone can make for home energy savings.

One shutter tested by Shurcliff stands out as being particularly simple and effective. It consists of foil-faced isocyanurate foam $\frac{1}{2}$-inch thick (Figure 16). This material is very lightweight—so light that dropping it on your foot, or, as Shurcliff says "on a baby's head," couldn't possibly do any harm.

To make the shutter, cut the foam board to size so that it overlaps

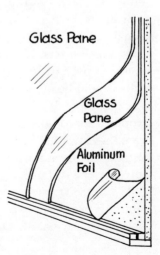

Figure 16 A foil-faced isocyanurate board makes a very effective thermal shutter.

the window sash a little. Seal the edges of the shutter using duct tape or any other reasonable technique so that the insulation itself is protected. If you want to, decorate the surface with fabric or paint. You will also have to attach a handle or cut a hole in the middle of the board to give you a purchase. A hand-sized hole makes no detectable difference in the shutter's thermal performance and does not interfere with stacking as a handle does.

Finally you need a way to hold the shutter in place on the window. The bottom edge can rest on the window sill or (if you're fitting the shutter to the upper window of a double-hung window) on top of the sash of the window below. The shutter can be held by nails, screws or pegs near the top, with just enough room to slip the shutter in place between the pins and the sash (Figure 17).

Another option is to install clips at the top of the sash to catch the top edge of the shutter and hold it to the sash (Figure 18). To ease installation and removal, the vertical tabs should be just barely long enough to hold the shutter in place.

A very low-cost alternative to the isocyanurate foam shutter is to use several layers of cardboard covered with aluminum foil on both sides. No fire testing has been done, so it is impossible to know how enthusiastic the National Fire Protection Association will be about this shutter; but it works, and you cannot beat its payback time.

When used on tight, double-glazed windows, all these shutters reduce the heat loss rate by 60 to 80 percent when the outside temperature is 30°F.

William Shurcliff has tested a wide variety of other window curtains and shutters. He found a single layer of 1-mil plastic sold as a window cover to be particularly ineffective (only 5 to 10 percent reduction in heat loss at 30°F.). Polyethylene is not a good window cover because it is transparent to infrared ("heat") radiation—a property that makes it very suitable for use as a window in an infrared spectrophotometer in a chemistry laboratory, and very unsuitable for use as a window in a house. The best that can be said for polyethylene is that it reduces infiltration through a leaky window.

If you are worried about plastic foam shutters because of their flammability (see Chapter IV for more on this), you can substitute rigid fiberglass board. Or you might do some carpentry and build

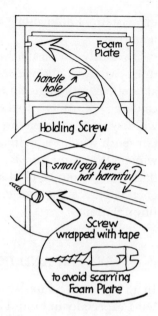

Figure 17 A simple mounting for a foam window shutter.

reflective shutters or shutters stuffed with fiberglass.[3] Obviously this is more time-consuming than using a foam board, and, thermally speaking, the result is not apt to be an improvement over a simple foam board. One solution is to use the isocyanurate shutter and at the same time invest in several smoke detectors for an early warn-

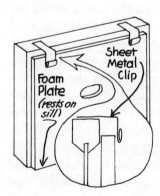

Figure 18 Another way of holding shutter against the window-frame.

ing system. Photovoltaic smoke detectors contain no radioactive materials.

Obviously an insulating shutter is more cost-effective than a storm window, and can eliminate the need for the storm window altogether, although not the need for good weatherstripping and caulking. Shutters can be left in place all winter long on some windows—for example, in bedrooms which are used only at night. On south windows, management of the shutters is all-important in winter. When the sun is out, the shutters should be off; when it's dark they should be in place. On cold and very cloudy days, it saves energy to leave the shutters on the windows and use a light bulb for illumination.

WATER AND INSULATION

In an insulated house, water and water vapor can become big problems, especially in humid and more extreme climates. Plastic foams are generally not terribly susceptible to damage by water (although urea formaldehyde does not stand up to high humidity); but with fiberglass, cellulose, and other fibrous insulants, great care must be taken during installation to seal water out. There are two sources of water that must be considered: water vapor and precipitation.

Water Vapor and Vapor Barriers. The difference in temperature that causes heat to flow through the thermal shell also drives moisture into the wall from the warm side toward the cold side. As the vapor moves through the insulation toward the cold side, its temperature drops. At some point condensation may occur, a phenomenon which depends both on temperature and humidity. The temperature at which condensation occurs is called the *dew point* temperature. In regions of high humidity the dew point temperature can be quite high. For example, along the Gulf Coast in Louisiana in summer the dew point temperature can be as high as 75°F.

Wet insulation creates huge problems. For one thing, wet insulation has a much lower R-value than the original dry insulation, so heat flow can increase dramatically. Also, if studs, sheathing, and other wood remain wet for a period of time, they may rot. Under a snow load rotten timbers collapse; it takes only a few years from the time of initial wetting. And finally, electrical fixtures in contact with wet insulation are a fire hazard.

Water vapor problems can be controlled with the use of a vapor barrier, a layer that is highly impervious to water vapor; it is placed on the *warm* side of the insulation. That is, on the inside of most buildings. Only air-conditioned buildings in very hot, humid climates must be protected from moisture penetration from the outside by vapor barriers.

In cold northern climates, a vapor barrier is always an indispensable feature of an insulated wooden wall. Even with plastic foam insulations that are largely impervious to water vapor, a separate polyethylene vapor barrier is often used to eliminate the possibility of any penetration by moisture. Also, vapor barriers provide an extra seal to prevent infiltration of cold air, so they do double duty.

In warmer climates, vapor barriers are not always used. There are three reasons for this: First, the temperature drop across the insulation is usually less, so that vapor migration is slower. Second, the smaller temperature drop across the wall is also less favorable to condensation. Third—and most important—where the interior of the house is subject to heating in winter and a significant amount of cooling in summer, the vapor barrier would be on the wrong side of the wall much of the time. And the only thing that is worse than no vapor barrier is a vapor barrier on the wrong side of the insulation. When it's on the cold side of the insulation, the barrier traps moisture inside the wall.

The effectiveness of a vapor barrier is measured in *perms*, obsolete units retained only in the United States and equal to one grain of water per square foot per hour per unit vapor pressure difference in inches of mercury.

The definition of a vapor barrier used to be anything with a perm rating of less than 1, but this is no longer adequate. A better approach is to consider the relative permeabilities of the outer and inner sides of the wall. Note in Table 4 that modern exterior plywood has a perm rating of less than 1, so that it is a vapor barrier by the old definition. Nevertheless, it would not do to rely on polystyrene foam as a vapor barrier on the inside with plywood on the outside. Water would migrate through the foam and very likely condense on the plywood, whose glue blocks further outward migration. That is why it is absolutely essential to use a polyethylene or aluminum foil vapor barrier on the inside over foam.

A rule of thumb in use among builders is that the ratio of permeabilities of outer and inner skins should be no less than 5. In the case of plywood and 6-mil polyethylene this ratio is 0.5/.05 or 10, whereas it's ½ for plywood and 1 inch of polystyrene. A hole ¼ inch in diameter in polyethylene causes a threefold increase in its perm rating, lowering the ratio from 10 to 3.3.[4] This explains why thoughtful builders take such care installing polyethylene vapor barriers and why they resist putting electrical outlets in outside walls. A careful builder will buy polyethylene in sheets big enough to cover an entire wall and thus avoid the necessity of joining smaller sheets. Note that the ratio of perm ratings for traditional building paper and polyethylene is 4/.05 or 80, well above 5.

Precipitation and Insulation. The other source of water that may get into your insulation is precipitation. High-quality construction should shed rain and snow effectively so that no water leaks into the

TABLE 4
Permeability to Moisture of Various Building Materials

Aluminum foil, 1-mil	0 perms
Polyethylene, 6 mil	.05
Polyethylene, 4 mil	.08
Kraft paper, foil-faced	.5
Exterior plywood, ½ inch	.5
Kraft paper, asphalt impregnated	1.0
Vinyl wallpaper	1.0
Exterior oil-based paint	1.0
Urethane, 1 inch	0.4–1.6
Extruded polystyrene, 1 inch	1.2–3.0
Beadboard polystyrene, 1 inch	2.0–5.8
15-pound asphalt-treated building paper	4.0
Urea formaldehyde, 1 inch	26
gypsum board, ⅜ inch	50

insulation. However, to be safe it is a good idea to make sure that water that does find its way into a wall, whether owing to careless construction or extreme weather conditions, is not trapped permanently.

The way to accomplish this is to ventilate the cold side of the insulation. As long as air can circulate freely, the insulation will eventually dry out. If such ventilation is not provided and the insulation does get wet, it will stay wet; the insulation will remain ineffective, and rotting of structural wood and corrosion of electrical fixtures will be vastly accelerated.

The walls of older houses with wooden siding generally are adequately ventilated without any special provisions. In newer construction, ventilation of walls is sometimes specially provided.

You must keep an eye on your walls and monitor what's going on inside. Peeling paint means that water vapor is trying to get out. The sun often dries out the south wall quite effectively, so look for signs of deterioration on the north wall first. Carpenter ants are a sure sign of moisture penetration. If the walls are constantly damp, plastic ventilating plugs can be installed. Another way to help dry out damp walls is to allow the paint to blister and flake off. Then, instead of putting more paint on, treat the wood with a stain that does not impede moisture migration in the least.

As long as moisture buildup within a wall is not really excessive, it is to some extent self-correcting. As the wall gets wet the R-value of the insulation drops, which results in faster heat loss; the temperature in the wall rises, and this in turn opposes further condensation and speeds evaporation.

Excessive Humidity. A good vapor barrier tends to keep household humidity high, both by preventing moisture from migrating into the walls and by stopping infiltration of outside air, which contains very little moisture in winter. This saves the energy required to humidify a house in winter, energy that is equivalent to 80 gallons of oil per year in extreme cases. Therefore, a vapor barrier is even more valuable than one might at first expect.

However, it is not good to allow the humidity to become too high in winter. If your vapor barriers and seals against infiltration are too good, the evaporation that occurs naturally in a house—3 to 10 gallons per day—can raise the humidity to an uncomfortable level and lead to condensation on any cold surface. For this reason, it is common not to use a vapor barrier below attic insulation in very tight houses. True, moisture rises into the attic, but it is then carried away through attic vents. This method of coping with excessive humidity where there is adequate attic ventilation is approved of by the National Association of Home Builders.[5] Their rule of thumb, based on many years of experience, is that a vapor barrier is needed in the ceiling below a well-ventilated attic only if the design temperature is less than $-20°F$.—that is, only in the coldest parts of the United States with 8000 or more degree-days of heating. Vapor barriers should, however, always be used below cathedral ceilings, because ventilation on the cold side is apt to be minimal.

To give a quantitative idea of what adequate attic ventilation is, we cite the FHA minimum property standards which call for a ratio of vent area to ceiling area of 1/150 in the absence of a vapor barrier in the ceiling and 1/300 in the presence of a vapor barrier with a perm rating of less than 1. One way to tell whether your attic has adequate ventilation is to look for rust on nails protruding through the roof. If the nails are rusty, there is too much moisture in the attic and the vents should be larger.

If you are retrofitting insulation into the attic floor in a house that tends to be too dry in winter, then do put a vapor barrier below it. This may be polyethylene, aluminum foil attached to batts or blankets, or a low-perm paint on the ceiling below. If the humidity then gets too high in winter, you will have to either use a dehumidifier or air out the house from time to time or reduce the rate of production of water vapor, by, for example, putting a vapor barrier down on the earthen floor in the basement or ventilating the bathroom.

Two last points about water vapor: (1.) If the cold side of your thermal shell has a very low permeability, e.g., asphalt shingles or vinyl or aluminum siding, it is especially important that adequate venting be provided so that moisture isn't trapped inside the wall; and (2.) When the cold side of the shell is fairly tight, any caulking should be done on the inside so that the wall is not prevented from breathing. You have to use your own judgment. Obviously, large cracks on the outside should be sealed so gross water penetration does not occur. On the other hand, in most houses much of the caulking should be done on the inside.

FIRE SAFETY AND INSULATION

An insulation job has the unusual characteristic that it can create fire hazards in your home, or, if it is done well, it can make your home more resistant to fire. In a burning house, heat and flames can travel rapidly through the wooden structure so that the entire house is soon involved. Proper construction with firestops between studs and fire-resistant surfaces will slow down the progress of a fire.

Insulation can contribute greatly to the ability of a section of the structure to resist penetration by fire. It is quite possible for one surface of an insulated wall to be burning fiercely, and the other side to be cool to the touch. Thus, insulation can give a family extra minutes to respond to a fire.

However, care must be taken when installing insulation in order that the fire hazard is not increased. Some insulating materials are totally non-flammable, such as vermiculite and perlite. Other materials are extremely difficult to ignite. Urea formaldehyde foams that are injected into walls are in this class. And finally there are materials which, although they do not burn by themselves easily, can readily contribute to a fire. In this group are most urethane and polystyrene foams.

Unless it is extremely combustion resistant, no insulating material should be left unsheathed. Plain gypsum board provides a good barrier to fire. In many homes, flammable insulation is sheathed with wood. This is less than ideal, although the owners are probably correct in believing that in case of fire the insulation is not what is going to be the major threat. The wood will be totally involved before the fire penetrates to the insulation.

Another potential fire-related danger from insulation is the possibility that insulation, being insulation, will trap heat inside a house during a fire and thereby increase the rate at which fire spreads. This is something that needs much more investigation. The few tests that have been done so far are in no way conclusive.

Fire hazards are best resolved in the same way you solve water-related problems: fix them before they cause trouble. Careful construction is the best way to make sure that you will never need to worry that your insulation is a fire hazard.

R-VALUES OF DIFFERENT WALL CONSTRUCTIONS

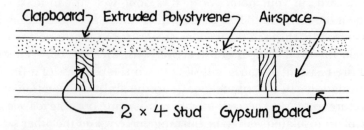

Figure 19 Calculation of total and effective R-values of a 2×4 stud wall sheathed in extruded polystyrene foam. This wall is energy efficient and easy to build. If the stud cavities are filled with brick, the heat storage capacity of the wall is increased tremendously. It is best to use a polyethylene vapor barrier between the gypsum board and studs, unless there are electrical fixtures in the wall, in which case the vapor barrier should go between the studs and the foam.

	R-Values in Framed Section	R-Values Associated with the Uninsulated Air Space
Outside air film (in 7½-mph wind)	.25	.25
Clapboard siding	.80	.80
2" extruded polystyrene (R = 5.5 per inch)	11.00	11.00
3⅝" studs (R = 1.25 per inch)	4.55	—
3⅝" air space	—	1.00
½" gypsum board	.45	.45
R_{tot} =	17.05	13.50

If the framing constitutes 20% of the total wall area, then the effective R-value of the wall,

$$R_{\text{eff}} = \frac{1}{\frac{.20}{17.05} + \frac{.80}{13.50}} = 14.09$$

Understanding Heat Flow in Buildings

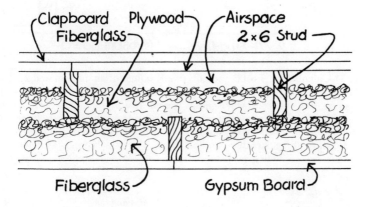

Figure 20 Calculation of total and effective R-values of a double 2×6 stud wall (24 inches o.c.) insulated with fiberglass. This is the type of insulation often used in new, super-insulated houses. Off-setting the studs blocks heat leaks through the studs. The 2-inch air space just inside the exterior sheathing provides ventilation to keep the fiberglass dry.

	R-Values in Outside Framed Section	R-Values in Inside Framed Section	R-Values in Insulated Section
Outside air film (in 7½-mph wind)	.25	.25	.25
Clapboard siding	.80	.80	.80
½" plywood (R=1.25 per inch)	.60	.60	.60
5⅝" stud (R=1.25 per inch)	7.05	—	—
2" air space	—	1.00	1.00
Fiberglass, R-11	—	11.00	11.00
5⅝" stud (R=1.25 per inch)	—	7.05	—
Fiberglass, R-19	19.00	—	19.00
½" gypsum board	.45	.45	.45
Inside air film	.70	.70	.70
R_{tot} =	28.85	21.85	33.80

If the exterior framing constitutes 15 percent of the total wall area, the interior framing constitutes 15 percent of the total wall area, and the balance is insulated, then the effective R-value of the wall,

$$R_{eff} = \frac{1}{\dfrac{.70}{33.80} + \dfrac{.15}{21.85} + \dfrac{.15}{28.85}} = 30.51$$

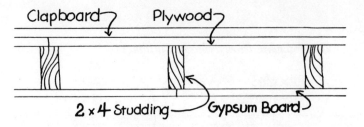

Figure 21 Calculation of total and effective R-values of an uninsulated 2×4 stud wall (16 inches o.c.). This archaic form of wall construction is still very common in older homes that were built when energy was cheap and insulants were not as good as they are today. Very often the exterior sheathing and siding are so loose-fitting that air circulation between the outside and the stud cavity virtually destroys the R-value of the siding, sheathing and air space. The calculation is as follows.

	R-Values in Framed Section	R-Values Associated with the Uninsulated Air Space
Outside air film (in 7½-mph wind)	.25	.25
Clapboard siding	.80	.80
⅜″ plywood (R=1.25 per inch)	.45	.45
3⅝″ air space	—	1.00
3⅝″ stud (R=1.25 per inch)	4.55	—
⅜″ gypsum board (R=.90 per inch)	.35	.35
Inside air film	.70	.70
$R_{tot} =$	7.10	3.55

If the framing constitutes 20 percent of the total wall area, then the effective R-value of the wall,

$$R_{eff} = \frac{1}{\frac{.20}{7.10} + \frac{.80}{3.55}} = 3.94$$

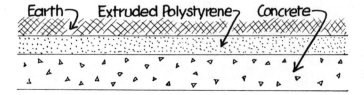

Figure 22 Calculation of total R-value of a poured-concrete cellar wall insulated with extruded polystyrene foam. Where 4 inches of foam are used on the upper part of the wall 2 may be used on the lower part. If the insulation were on the *inside*, it would require protection from fire, and the moderating effect of the mass of concrete on the temperature would be lost.

	R-Values
Earth (worst case)	0
4″ extruded polystyrene (R=5.50 per inch)	22.00
8″ poured concrete (R=.10 per inch)	.80
Inside air film	.70
$R_{tot} =$	23.50

Understanding Heat Flow in Buildings

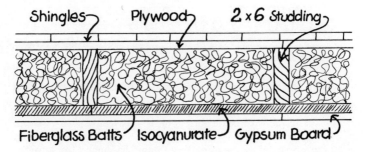

Figure 23 Calculation of total and effective R-values of an insulated 2×6 stud wall with high–R-value interior sheathing of isocyanurate foam. The interior foam layer improves the R-value of the wall and also its vapor barrier. Foam and gypsum board can be obtained as a unit.

	R-Values in Framed Section	R-Values in Insulated Section
Outside air film (in 7½-mph wind)	.25	.25
Wooden Shingles	.85	.85
½″ plywood (R=1.25 per inch)	.60	.60
5⅝″ studs (R=1.25 per inch)	7.05	—
Fiberglass batts, R-19	—	19.00
1″ isocyanurate foam	7.50	7.50
½″ gypsum board	.45	.45
Inside air film	.70	.70
R_{tot} =	17.40	29.35

If the framing constitutes 15 percent of the total wall area, then the effective R-value of the wall,

$$R_{eff} = \frac{1}{\frac{.15}{17.40} + \frac{.85}{29.35}} = 26.60$$

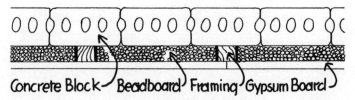

Figure 24 Calculation of total and effective R-values of a concrete block wall insulated on the inside with beadboard. This is a reasonable way to insulate a damp basement wall from the inside, since the foam is not destroyed by dampness. However, there is no good way of insulating a damp basement. Insulating on the inside may increase the danger of frost damage to the blocks below grade.

	R-Values in Framed Section	R-Values in Insulated Section
Outside air film (in 10-mph wind)	.25	.25
8" concrete block	1.20	1.20
1¾" beadboard (R=4.4 per inch)	—	7.70
1¾" framing (R=1.25 per inch)	2.20	—
½" gypsum board (R=.90 per inch)	.45	.45
Inside air film	.70	.70
R_{tot} =	4.80	10.30

If the framing constitutes 15% of the total wall area, then the effective R-value of the wall,

$$R_{eff} = \frac{1}{\frac{.15}{4.80} + \frac{.85}{10.30}} = 8.79$$

NOTES TO CHAPTER II

1. G. A. Tsongas, R. S. Carr and C. D. Stultz, "An Experimental Study of Solar Heating Effects on Wall Insulation Performance." Paper delivered at the Third National Passive Solar Conference, San José, California, 1979, page 723.
2. William Shurcliff, "Heat Loss through Double-Glazed Windows with Thermal Shutters and Shades." Unpublished. Dr. Shurcliff is a physics researcher who may be contacted at 19 Appleton St., Cambridge, Massachusetts 02138.
3. For reflective shutters, see C. G. Wing in *Organic Gardening,* November 1977, page 121. For making foam shutters attractive, see J. Ruttle in *Organic Gardening,* December 1978, page 83.
4. W. F. Carpenter, "Vapor Barriers for Buildings Having High Atmospheric Moisture Conditions," Maine Agricultural Experiment Station Bulletin 623.
5. *Insulation Manual.* National Association of Home Builders Research Foundation, Second Edition, 1979, page 45.

CHAPTER III

Weatherstripping and Caulking

Roger Albright

Insulating in general is the process of placing a temperature barrier between the indoors and the outdoors. As suggested in the preceding chapters, this barrier is only as good as it is tight and free of leaks, cracks and holes. The money, time and effort you put into proper insulation can be wasted entirely if your structure still has holes where cold winds can penetrate. This chapter gets specific about hole-plugging.

For example, consider your front door. On an average day it may be opened and closed ten times. That makes more than 3000 slams a year. No matter how well built and tightly fitted to begin with, that door may eventually not fit as well as it did at the factory.

Let's suppose, then, that the fair wear and tear has opened a gap around your front door that averages only $\frac{1}{16}$ inch wide on the two sides and the top. That may not be worth bothering about, but it is the equivalent of a 3 × 4-inch hole—plenty big enough to put your fist through. If you had a hole in your wall that big you'd fill it before winter came. If you have a $\frac{1}{16}$ inch gap around your front door, you *do* have a big hole in your wall that needs filling.

Suppose we cut that gap in half. Make it just $\frac{1}{32}$ inch, or about what you could slip a knifeblade into; that's still the equivalent of a hole 3 inches by 2. You'd be sure at least to stuff a rag in it the first time a cold wind was blowing. Caulking and weatherstripping is more work than stuffing a rag in a hole, but you can be sure it's just as necessary if holding down your fuel bills is a consideration.

Even a brand-new home will likely have some cracks, gaps, and apertures that need sealing, and an older home is likely to be about as tight as grandpa's favorite slippers unless it has had careful and continuing attention.

Continuing. There's a key word. Another year means another 3000 slams for the faithful front door and time for another checkup. All the way through this chapter you'll be reading tips and suggestions that probably should be part of a yearly schedule of checklist items. In the caulking and weatherstripping category there are few projects that are taken care of once and for all.

Yes, the front door should probably be checked every year, and no kind of weatherstripping there should be expected to last more than 5 years. A good job around a window pane may last for 15 years, but if it's exposed to sunlight on a south wall, or if you're near a highway with the constant rumble of traffic, that putty or glazing compound may start cracking within a year. A severely cold winter can open up cracks in masonry in a matter of weeks, even though you just completed a thorough pointing job.

The annual checklist to be reviewed, say, every August, is a fine idea. There's still some pleasant weather ahead for outdoor work, and who knows?—maybe you'll discover to your relief that your checkup reveals only a few little items to tend to.

CAULKING WINDOWS

In conventional construction, a properly installed window won't leak cold air; right? Not necessarily.

In the building of a house, a window is a large hole in the wall with a window frame stuck in it like a plug. If you're lucky the plug is a close fit all the way around, but only if you're lucky. More likely, the window frame is a very near approximate fit to the hole that was cut in the wall to take it, and this is more likely to be true in a new house than an old one.

With construction labor costs what they are today, the contractor can figure on saving a buck if the hole is just a little too big so the window frame will fit in and square up properly in minimum time.

Weatherstripping and Caulking

As Figure 25 shows, the frame is designed for just such treatment, but it leaves you with potential breeze gaps. After the window is installed, the finish siding is applied right up to the framing. Here, too, the haste of modern construction methods offers the opportunity for little gaps in a new house. Over time, the natural shrinkage of lumber also causes gaps around window frames.

For all these reasons, window frames are an item for your checklist.

Where the siding meets the window frame on all four sides is your place for a bead of caulk. If you can see that there's no visible gap, you're OK; but if you can see a dark line, or if you can wedge in a knifeblade, then you're about to solve a problem with your caulking gun.

For the amount of work you'll be doing, you don't need a professional caulking gun. An inexpensive tool from your neighborhood hardware store will do the trick adequately.

You'll have a choice of caulking compounds. Check the label on the tube. If the compound has a lead base, don't buy it: the lead is toxic and offers no special advantages. The cheapest caulk will have an oil or resin base. You'll put less money down on the counter, but you'll also be doing the job over again sooner, because the cheaper caulks tend to dry out faster, crack, and shrink.

The latex and butyl caulks are more durable and, unless you plan to move soon, are worth the extra money because they last longer.

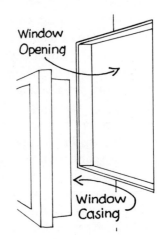

Figure 25 Pre-fabricated windows are plugged into the wall—sometimes loosely.

At the top of the line are the elastomeric caulks: the ones with a silicone, polysulfide or polyurethane base. They'll be most expensive and most durable. Wait for a fix-up sale and buy a bunch of tubes. It will be a good investment.

Recently a new caulking product has appeared on the market: foam caulk in an aerosol can. This writer has had no personal experience with the product; but on the basis of early tests recorded in *Consumer Reports* (April 1979) it appears to be an idea whose time has not yet come.

The Consumers Union testing indicated that applying the foam caulk takes a very skilled hand to achieve a smooth result; the product air-hardens, so there's no going back to touch up a sloppy job. Also, it adheres poorly to many surfaces and must be painted if it is to have durability in outdoor applications. In addition, the product has a tendency to stick in the aerosol valve, making it impossible to open.

On the basis of the information at hand, we'd recommend against trying the foam caulks, in spite of some apparent advantages. (Don't say we didn't warn you.)

How much caulking compound do you need? Well, in normal work a tube of caulk will do about two windows or two doors. That's normal work. But maybe you have a real problem. Your visual checkup tells you that you have some cracks around the window frames you can almost stick a finger into. You can see yourself using two tubes of caulk on a single window. Here's the bad news: With big cracks like that you can stand in one spot and crank a whole tube of caulk into one hole, because it's just dripping down between the inside and outside walls, and doing no good at all.

Big cracks and gaps must be filled with something less expensive than caulking compound before you pick up your gun. A natural expedient is to stuff that big crack with shredded newspapers. This will work, but you should know that you're adding a fire hazard to your home. That paper can eventually become so dry that the smallest spark would cause it to burst into flames. Not good.

A better choice is to stuff some fiberglass insulation into that crack. Tamp it in firmly with a screwdriver blade, then run your bead of caulk. If that isn't handy, then your hardware store will likely have some of the oakum used by plumbers and boat builders, which is an excellent alternative at modest expense. Another alternative is the polyfill (made of shredded foam rubber scraps) that is used in cushions and pillows. This makes a good hole-filler as a backup for caulk and doesn't present the fire hazard you take on with newspaper stock.

Using the caulking gun will take a little practice. Your first few runs may be effective but a little bumpy. If this is your first experience, we suggest starting on the back of the house to do your learning, where the visibility is at a minimum. By the time you get around front, you should be laying down a smooth bead.

For those early, bumpy runs, you can smooth the job with your fingertip, or with a screwdriver blade. It takes an extra minute or two, but you'll be happier with the result. This smoothing technique will also tend to assure that the caulk is firmly seated in the crack, not just sitting on the surface, soon to peel away.

Before you lay down a bead of caulk, clean the crack. If you don't, dirt, paint scales, old, cracked caulk and spiderwebs will keep your new bead of caulk from making a tight seal. Scrape with a spatula or screwdriver blade or pass over the crack firmly with a wire brush, and your job will be many times more satisfactory.

Incidentally, tubed caulking compound comes in colors to match your paint job. Look around. If the first hardware store doesn't have what you want, go shopping. You may have to put a lot of work in the job before you get done, so it might as well look good. You want to be able to stand back when you're finished and say "Aaah, nifty."

CAULKING DOORS

The process for caulking doors is about the same as for windows, except that a door frame may present a different set of problems. With opening and closing, opening and closing, a door frame can begin to get loose in its moorings. If that's the case, then the best caulking job may not do much good.

At the time of construction, a door is installed in much the same way as a window. A hole is cut in the wall, framing studs are nailed in place, and the door frame is fitted in like a plug and nailed to the studs. For a door it's even less likely than for a window that the hole in the wall was cut to the exact dimensions of the door frame. Because fitting a door is trickier than fitting a window, a certain amount of maneuverability is needed for the installer.

Over the years, that door frame can actually come loose from the house a little. To find out, open the door, grasp the frame, and try to shake or wiggle it. If there is the slightest amount of play, then you need to firm things up before you do the caulking.

The line of nails holding the door frame in place is probably under the strip or panel of the frame against which the door strikes. To firm a wobbly door frame you can either pound spikes right through the striker panel to the studs, or you can pry loose the panel, pound the spikes through, and put the panel back in place. No matter how careful you are, with either method you probably will have a repaint job ahead of you by the time you get finished.

With that door frame standing firmly, your caulking task is just what it was for the windows except, of course, for the sill. We'll get to that a little later when we're dealing with weatherstripping.

CAULKING FOUNDATION SILLS

The foundation of your house may be poured concrete, cinderblocks, or mortared stone. Whatever it is, what sits on top of that foundation is a sill or plate from which arise the vertical studs of the outside wall, as in Figure 26. The masonry and wood expand and contract at different rates, and it's unlikely that they fit tight to each other in the first place. What that means is a potential crack between foundation and sill all around the perimeter of your house.

Figure 26 Often there are gaps between foundation and sill.

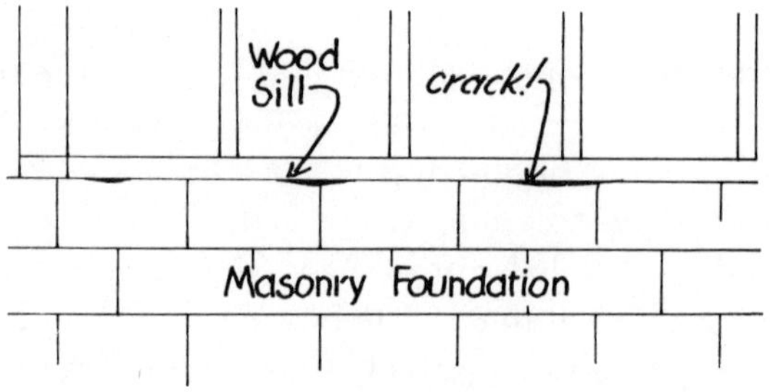

Weatherstripping and Caulking

The way the siding was installed, that gap may not be visible from the outside, but it's there just the same—a nasty little breeze entry. If you can't get at it from the outside, maybe you'll need to spot that juncture from down in the cellar or crawlspace, but in any event you can't ignore it.

When you tackle this job you might just as well decide before you start that you'll need some fiberglass or oakum, because there will almost certainly be at least a few spots where the gap is $\frac{1}{4}$-inch wide or wider, and it will need to be stuffed with filler before you caulk.

You'll know what an important job you did when you get finished and discover that you used at least four tubes of caulk to fill a crack that was the equivalent of a hole big enough for the cat to run through without missing a stride.

Your attached garage was likely designed and constructed this same way, and you'll do yourself a favor by caulking the foundation/sill crack in the garage, too. It serves as a temperature buffer for the wall of the house it's on, and in any case its best function is to minimize the chill on your car overnight. It will do its job best if the foundation/sill crack is filled.

CAULKING THE CORNERS

Here's an easy one.

Chances are that at each corner of your house there are vertical trim boards covering the joining of whatever siding was used, like those in Figure 27. If there are wings on the house you'll have both inside and outside corners with those vertical strips.

At the edge of each strip is a potential breeze-catcher where cold

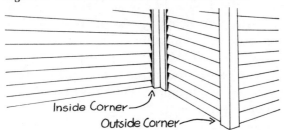

Figure 27 Look for air leaks around vertical trim.

air can sneak in just as it does around a window frame. Because the builder may have seen these strips as principally cosmetic they may not be nailed down as tightly as a door frame or window frame. If you can stick a screwdriver under the edge of the strip and wiggle it a little, a few more nails to bind it down firmly may be in order.

With the corner boards nailed down securely, a bead of caulk from top to bottom is a relatively simple task that will keep a lot of cold air outside where it belongs.

CAULKING THE MISCELLANIES

After the basic matters already covered, the most immediately visible of the miscellanies to be caulked is the juncture where your siding meets an outside chimney. The masonry of your chimney and the house siding are two different building materials that expand and contract at different rates, which is tough on whatever was used to seal the joint between them. If the material originally chosen was hard mortar, and it was applied more than a year or so ago, you almost certainly have a job to do, because the expansion and contraction has long since cracked that mortar to uselessness.

Clean the crack, stuff the bigger gaps with filler, then finish with a better grade of caulking compound. The better grade is most important here, because you need a seal with a degree of flexibility, rather than something that will dry hard and then crack.

Now that you've gotten this far, the principle at work in caulking is fully evident to you, as are the techniques for dealing with the problems. The rest of the miscellanies will pop out at you with just a simple eyeball inspection: outside water faucets and electrical outlets, outside electric meters or electrical entryways, dryer vents, junctures around porches, toolsheds, greenhouses—any break in the outside wall needs to be sealed against the infiltration of cold air.

There's one more very important point that must be mentioned. You're all done. You step back and look at your job. You see those neat new beads of caulk around the windows and doors. You feel good about your accomplishment. You've hit every point. Or have you? What about the siding itself?

Many kinds of house siding, and especially lap siding, will warp,

Weatherstripping and Caulking 65

crack and shrink with age. Particularly with an older house that hasn't been painted as often as it should have been, there can be entryways for cold air right in the siding itself. While you have your caulking gun at the ready and your ladder handy, take a close reading of the wall itself and fill the cracks that may have appeared.

It's a chore, but you can take comfort in the fact that every crack you fill confers a double benefit: you minimize the infiltration of cold air to save on your fuel bill, and you arrest the further deterioration of that siding, forestalling even more serious problems in the years ahead.

CAULKING STORM DOORS AND STORM WINDOWS

Putting up storm doors and storm windows can be an expensive proposition. You, or whoever had them installed, probably shopped around a little to get the best price. That "best price" may have meant that an important step in installation was skipped.

Speaking now of permanent storm window installations, the hasty contractor may have just put the windows and doors in place and screwed them down as tightly as he could. The step he skipped was putting down a bead of caulk between the new installation and the house casing before tightening down (see Figure 28). If that happened, chances are you don't have as tight a seal as you should,

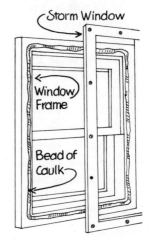

Figure 28 If storm windows are not going to be taken off in summer, they can be sealed with a bead of caulk.

and you're not getting full benefit from your storm doors and windows.

Take them all down? Yup. Because your patience may wear a little thin with this job, start on the cold north wall and get that done first, then get the windows and doors that face the prevailing westerly winds. That will be approximately half the job for much more than half of the potential benefit.

There's a reality to consider while you're doing this. Odds are that there are some windows in your house that are never opened. They are regular casement or double-hung sash that can be opened, but in truth they are useful only for letting in light or giving aspect on a view.

If you're prepared to say that a window doesn't ever need to be opened, because your experience tells you so, then seal it shut. Run a bead of caulk all the way around the window frame, and along the middle where the two windows meet in a double-hung sash, before you put the storm window back in place.

Without pondering, you can consider it a mandate to give this treatment to all fixed windows, like the picture window in the living room.

If, while you're doing this, you discover that a window you plan to seal rattles a little in its frame, then nail it firmly in place, stuff the resulting crack with filler, and be sure everything's firm before you caulk.

With this technique you still have all the light and visibility the window ever had to offer, but you have thwarted a certified money-loser. Incidentally, if you are just at the point of installing storm windows, those fixed windows won't need the expensive combination sash. Just simple windows installed in a bead of caulk will do it.

There's a point of controversy here. In some climates and with some kinds of construction, moisture will gather between the windows and storm windows, making it necessary to have drain holes, or "weeper" holes, at windowsill level to allow moisture to evaporate (Figure 29). The only way you'll know about this is from the actual performance of your house. Accumulated moisture can cause wood rot, which you don't want. If you have such moisture, then the caulk seal at the sill will need to be broken to allow the moisture to escape.

Weatherstripping and Caulking

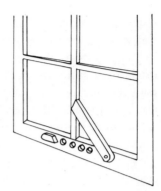

Figure 29 A storm window with "weeper holes" at bottom for ventilation.

Many factory-built storm windows are designed with weeper gaps, and you may want to leave them open when the windows are installed, but if you don't have a moisture problem, then the complete seal is most advantageous.

MASONRY POINTING

You may have already become acquainted with the business of "pointing" masonry when the first brick or two fell from the top of your chimney, or the first few cracks showed up in your front steps. "Pointing" is the term used by workers in brick and stone for repairing the cement joints that hold together the solid materials.

For the purposes of the subject under discussion here, we'll be looking first of all at cracks in the foundation, and secondly at cracks in a masonry facing. The pointing we'll be doing will add virtually nothing to structural strength. Its purpose is to fill cracks to keep the weather out. If you have serious structural cracks, you probably need to talk with a professional mason before your foundation starts to let you down. Literally.

Just as is true where wood doesn't quite meet wood, cold will infiltrate where stone doesn't quite meet stone, brick doesn't quite meet brick, concrete blocks have separated, or a poured foundation has cracked. When you fill those gaps you will be adding years to the life of your structure as well as keeping out the cold, because as moisture gathers in a masonry crack and then freezes, the expanding ice makes the gap even bigger and steadily weakens the structure year by year.

Because you'll be using a relatively small amount of repair material, a sack of ready-mix from the hardware store will do nicely. You can rent a portable concrete mixer, but you don't need it. Mix in modest amounts in an old bucket, following the directions on the package. Ready-mix for pointing "sets up" (hardens) fairly fast, so don't mix more than you'll be using in about 15 minutes. The mix will adhere to the sides of the bucket and be almost impossible to clean out.

When the package says "mix thoroughly" it means "mix thoroughly," but keep your hands out of the mix. Mortar mix will take your skin off. Mix with a long sturdy stick, and if you get any on your skin, rinse promptly at that handy outdoor faucet. The long stick is to minimize the possibility of breathing damaging mortar dust. You're mixing outdoors, so you won't be flushing mortar mix down your plumbing drains. Standard household plumbing doesn't take kindly to ready-mix, and it's just plain hell if the stuff sets up in an inaccessible elbow.

The only tool you really need is a mason's trowel. Because this will be a tool you won't use much, you won't need an expensive one. Just a plain, cheap trowel will do, if it has a little flex.

Most professional masons also use what they call a finishing trowel to hand-carry the mortar to the spot where they are working. If you have one handy, it's handy; if you don't, don't buy one. A scrap of board will do nicely for carrying lumps of mortar around. Professional masons also have a selection of tools for smoothing mortar. They're handy, too, but not essential. You can smooth and sculpt the mortar you have pressed into a crack with a screwdriver blade or the end of a dowel rod or the unsharpened end of a pencil.

Now we're at the crack. You've mixed your mortar in the bucket. You've scooped some out with your trowel onto a board for carrying it. You're ready. Here's the trick. You want to force as much mortar into that crack as you can. Just wiping a little across the surface may make the crack disappear for awhile, but you'll be doing the job again in a matter of weeks, or sooner. Pushing the mortar into the crack, you are likelier to make a permanent bond with the irregular surfaces inside for a repair that can last for years.

You didn't clean the crack first? That explains why the mortar fell out about as fast as you put it in. A wire brush or whisk broom will help to get the crumbs and cobwebs out of the crack. Further, if you wet the brush while you're cleaning so the crack is moist when you force the mortar in, you have an even better chance of getting a permanent bond as the repair sets up.

Pointing a brick facing or a concrete block foundation wall, the color of your mortar may be a factor. The ready-mix from the store is likely "natural" in color, which means that it will be a very light gray when it dries. If you need a darker color to match work that was done in the past, ask for some carbon black at the hardware store. You stir in this color additive as you mix the mortar. Master masons get paid a lot of money because they've learned how to get precisely the color they want without measuring. It's one of the secrets of their trade. You'll get close by the trial-and-error method. Just remember that the mortar mix is darker when it's wet than it will be when it's dry.

Applying wallpaper, you're working from the top down. Usually when you paint you're working from the top down. The general rule for masonry pointing is the opposite: from the bottom up. You want the mortar you lay first to support the mortar that follows.

But say you're confronted with a really big crack, or a piece that is altogether chipped away and missing. Top down or bottom up, neither is going to work because the mortar will fall out as fast as you stuff it in. So you don't frustrate yourself trying to fill a crack when what you're actually doing is rebuilding a wall. That's a larger job and requires what a mason would call "forms."

Don't be intimidated. Just a bunch of scrap boards will do. As Figure 30 portrays, you fill from the bottom up, adding boards as you rise. A corner is a little trickier, but the same principles apply. It's not too difficult, although it takes a careful and experienced craftsman to avoid the patchwork look in the end result. Again, that's why master masons get paid so well.

As you finish a section of work, it's a good idea to sponge off the surface before the mortar sets. You may have slopped a little onto the bricks, stone, or blocks, and the job will look better if you wipe away that excess. If you wait too long that excess will only be removable with a hammer and cold chisel.

Here's a point we made once before, but it's worth mentioning again. Where masonry meets another building material—usually

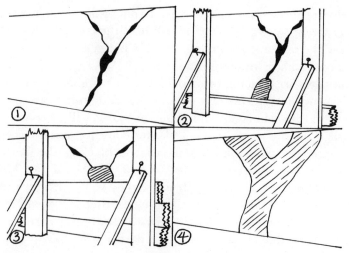

Figure 30 Steps in patching a major crack in a masonry wall.

wood—mortar mix is an altogether unsatisfactory joiner. Each of the materials expands and contracts at a different rate as the temperature varies; each has a different index of flexibility. If there is a crack where masonry meets another material, you need a caulking compound, as described earlier in this chapter. Stuffing mortar mix into that kind of crack is a waste of time and temper.

When the job is done, rinse the mortar bucket out thoroughly, but not down a house drain; clean your trowel and dry it (or it will rust), and take a bath or shower just to be sure there aren't any residues of mortar dust on you that might cause irritation. In about 24 hours, when the work has completely dried, you'll be able to see what a good job you did and know that each crack filled will be a certified fuel saver.

KINDS OF WEATHERSTRIPPING

Let's start with a quick survey of the various kinds of weatherstripping materials available on the market (see Figures 31-39).

Probably the least expensive—also the least satisfactory—is the tubular gasket strip, usually sold in rolls. This is essentially a plastic tube with a nailing flange, the tube sometimes stuffed with sponge

Weatherstripping and Caulking

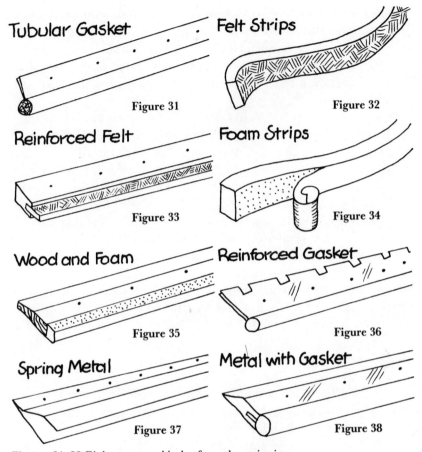

Figures 31–38 Eight common kinds of weatherstripping.

rubber or polyurethane, sometimes hollow. There are two problems with this. It is hard to get a really firm fit when installing it, and it wears out too fast.

Next cheapest, and not much better functionally, are strips of nonreinforced felt. It's easier to install than the tubular gasket stuff, but it wears fast, and isn't suited to a situation where the gap is of uneven widths along its length.

Reinforced felt strips cost more and aren't much better, because the material still wears out fast, is subject to weather erosion factors like mildew, and is somewhat difficult to install satisfactorily.

You can buy rolls of vinyl foam strips with an adhesive backing in various widths and thicknesses, but these, too, wear out fairly quickly. They are better for filling gaps of nonuniform width but can endure very little abrasion. Consumers Union ranks foam strips in this descending order: vinyl, neoprene, sponge rubber, polyurethane.

The next step up is wooden strips with a plastic foam or tubular gaskets glued to them. They install with relative ease, don't look too unsightly, will cover nonuniform gaps, and will install with either nails or screws. The problem is that the two materials used are sufficiently dissimilar that apparently the adhesive is not yet available to hold them together. In other words, this stuff looks good when you put it up, but the foam or gasket will separate from the wood in too short a time, rendering the installation useless.

Not as good looking, and less expensive than wood strips, but more satisfactory, is the metal-reinforced gasket. This is the tubular gasket noted above with a metal (usually aluminum) reinforcing strip to make possible a much more firm installation. It is still subject to early wear, but will last a lot longer than its nonreinforced cousin. It has enough flexibility to cover nonuniform gaps and—all told—is probably the best of the cheapies.

Just a step up in price, and a big jump up in durability, are the flat metal strips of brass or aluminum that make their seal by spring tension. In place, these are almost invisible; they are relatively easy to install and can last for years. The major disadvantage of the spring strips is that they are just not suitable if the gap varies in width from end to end, as it is very likely to do. To say it another way, this is a first choice in new doors and some windows, but fitting to older problems the spring strips may not do the job.

Now we're taking a considerable jump up in price to consider the aluminum strip fitted with a neoprene or rubber gasket. We're close to the ideal here. This has enough flexibility and cushion that it will work with nonuniform gaps, it isn't difficult to install, it scores well on durability, it doesn't look too bad. When you decide to get up out of the cheapies, this may be the best bet.

Before we go on, there's a problem to spotlight in connection with the rigid strips, when they are used on a door. Metal strips will be

Weatherstripping and Caulking

predrilled for nails, or, better, screws. Plastic strips will also be predrilled. That's good, because it will save you a lot of work, even if you have an electric drill. Some of those strips will be drilled with holes, some with slots. The idea behind the slots is that the strip is then adjustable, which seems like a good idea. It isn't. If the strips are adjustable, they will start adjusting themselves every time the door slams. Soon they won't be fitting tightly at all, because the slamming door has pushed them out of position. You'll adjust them, if you think of it, adjust them again, adjust them again, and after about the fifth time you'll discover that the screw isn't holding any more down there at the bottom of the door. Go for screw holes, not slots.

There are several kinds of formed metal weatherstrips for doors that make an interlock with strips on both door and frame (Figure 39). They are very efficient, very durable, and some are almost invisible. The most satisfactory of these work best on doors that fit pretty well to start with, all of them require a skilled carpenter for installation, and all are probably more expensive than the results can justify.

You may have seen an ad for a very expensive door kit with a flexible brass curve that installs with a pressure-sensitive adhesive (Figure 40). No nails, no screws, no muss, no fuss, no bother. Not much good, either. If the gap to be closed isn't essentially uniform from end to end the metal won't make a seal at all. Bump it going through with a piece of furniture and the installation is ruined. Since the obvious place for such an installation is an outside door, in due time the temperature changes are going to get to that adhe-

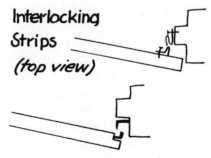

Figure 39 Interlocking metal strips for giving a door a tight seal.

sive, and there are no options for fixing it with nails or screws. Skip this one.

WEATHERSTRIPPING ENTRY DOORS

Let's take the toughest problem first. If your door sill is badly worn, you have a problem that will require some doing to resolve. Most hardware stores and lumber yards will sell you any one of several kinds of strips for the bottom of an outside door, but if the sill is visibly worn down, none of them will work effectively.

There are two options. The lumber yard will have oak planking specially milled for door sills. This is good. For heaven's sake, don't make your own sill out of a pine 2×6 you found in the garage. That won't last more than a few months, and a considerable effort will have been wasted. The other option is a metal door sill. This will be more expensive than the oak, but, properly installed, it will last a lot longer.

Assuming a reasonably level door sill, there are four kinds of strips to make the bottom of the door weathertight.

Probably the most satisfactory, also the most expensive, is apt to be beyond the skill of the average home craftsman. It involves one extruded metal strip that screws to the sill, and another extruded strip that screws to the bottom of the door, as in Figure 41. When the door is closed, they interlock. Nifty. The installation, though, must be quite precise to make this work; you'll go crazy trying to install it on a worn sill, and the strip screwed to the sill is subject to wear and clogging with wintertime debris, which negates the seal.

Like the interlocking metal strips, the gasket threshold in Figure 42 will probably demand that a small amount be trimmed from the bottom of the door, preferably on a bevel this time, which is in itself a trick for the skillful. It also has the same inherent problem as the interlock: you'll be walking on the seal as you go in and out, so it will wear out. With some designs the gasket is replaceable, but that's a nuisance. This one will also leave you crying if the doorsill is worn, but you can shim it up in the worn place.

Putting on a door shoe, shown in Figure 43, means taking off the door, even as it did with the two preceding, but that's an easy assignment you would have to do to install weatherstripping anyhow.

Weatherstripping and Caulking

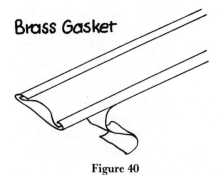

Figure 40

Figures 40–44 Five common threshold gaskets.

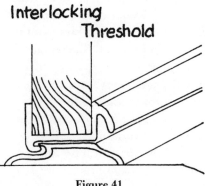

Figure 41

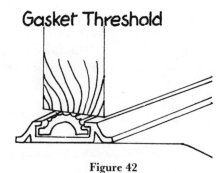

Figure 42

Figure 43

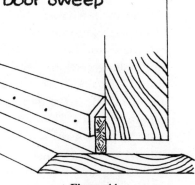

Figure 44

A door shoe is something like a threshold gasket upside down, with the advantage that—because it's on the bottom of the door—you're not walking on it. With a level sill and enough cut off the bottom of the door to install the gasket properly, this one will look good, perform effectively, and last quite a while.

The easiest to install is also the least expensive, and in some cases the most effective. It's called the sweep; it is a metal-reinforced strip of plastic (best), rubber (not bad) or felt (will wear out fast) that screws to the bottom of the door (Figure 44). It is subject to the most abrasion of any of the options, and so will wear out the soonest, and it is the most visible and least attractive. However, there is no need to trim the door itself to make the installation, which is a considerable advantage, and it will probably do the best job of dealing with an uneven sill. If you have a shag carpeting in your entryway (why? because you're in love with constant housework) the sweep may drag on your carpet; you may want one of the sweeps that lift automatically when the door is opened and fit down in place again when the door is closed. Someone has thought of everything.

As you can see, there is no perfect solution to sealing the bottom of an entry door, but, for you, some answers are going to be better than others. Assess your situation and act accordingly.

For the door itself on the two sides and the top you have another situation to assess. Assuming that the door opens in, go outside, close the door, and then give it a good push at the top above the handle, and then at the bottom below the handle. If there is considerable play, any of the rigid strips for sealing around the door may let you down. In fact, the first step may be to reposition the striker plate at the latch to get a firmer closure, and then test again for play at top and bottom.

If there is still considerable play, it may be that neither the metal-reinforced gasket strip nor the rigid extruded metal with gasket will do the job. You have a bad door and your best bet may be either spring metal, if the gap is fairly uniform from top to bottom, or adhesive-backed foam affixed to the door jamb.

If the play of the closed door is minimal, then one of the gasket options will do the job. With the door still closed, press the gasket strip firmly into position at the top and either screw or nail it into place. Then do the same at the bottom. What you're trying for is to minimize the play that exists, to be sure that the gasket will fit securely over a considerable time. Screws are best for fastening, but you'll find that working with a screwdriver in this position is awkward. Start the screw with the door closed, then open the door to finish driving it in place. That first vertical strip is the most important. The top and the hinge side will be easier, with lots of opening and closing of the door.

You'll follow the same process with wood strips if you disregard our advice and decide to use them for cosmetic reasons.

Now that the front door is snug and tight, don't forget that each door opening onto an unheated area needs the same treatment. The door to an unheated cellar, the door to the garage, the door to an outer breezeway, the back door—all are potential breeze sneakers and should be checked and stripped.

If you have a truly ill-fitting door and lack the inclination to replace it right now, you may need to consider a new door jamb all around. This isn't likely to be a beautiful alternative, or even too practical on the sill, but it will work. Figure 45 will give you an idea of how it's done.

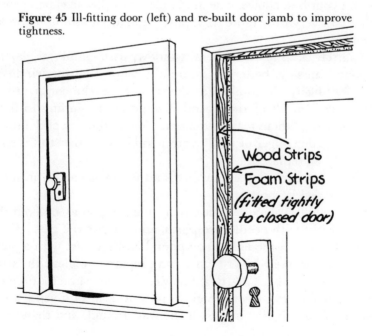

Figure 45 Ill-fitting door (left) and re-built door jamb to improve tightness.

CHECKING THE WINDOWS

It isn't prudent or necessary to assume that the windows in your house are going to need weatherstripping. Windows get nowhere near the action-wear that you give to entry doors, so just because you found that the doors needed work doesn't mean that the windows are also betraying you. If you can't jiggle them in their frames; if when you close them the latches seem to secure them snugly; if your eyeball inspection reveals no gaping cracks; if you haven't seen the curtains moving in the breeze since you did the exterior caulking—then your windows may be OK.

Before you set out to do window weatherstripping, check to discover whether your leakage is around the panes rather than around the frames. The best of glazing compounds will ultimately dry out and crack away, leaving apertures where warm air can escape and cold air sneak in.

Glazing compound. That's what used to be called putty. You can still buy putty, but regular putty dries out very fast, comparatively, and isn't your best choice. The newer elastic or plastic glazing compounds cost very little more than putty, and are substantially more durable.

If your checkup reveals several windows with chunks of glazing compound missing, better do a careful review of all the windows. They were likely all installed at the same time; the compound is equally aged on all of them, and if some are beginning to flake away, then all of them may be doing it soon. If you can peel out the sealer with your finger or easily with a knifeblade, the stuff is shot and should be replaced.

Let's outline the complete process, and then you can do as much of it as you feel is necessary.

At best, you'll strip out all the old glazing compound, pull out the glazier's points with needlenose pliers, and remove the glass. Then you'll clean out all the flaked compound and sawdust with a jackknife, chisel, or screwdriver blade, and prepare to reglaze the window (put the glass back in). Right now comes the reason for doing it this way. Before you reset the glass, press a bead of glazing compound into the window frame all the way around, and then press

Figure 46 Glazier's points.

the glass into that bead of mastic. This makes an additional cushion to protect the glass from breakage; it assures the tightest possible seal against infiltration, even with a wind blowing; and that inner bead is protected from the elements and will stay effective a long time.

The next step is the glazier's points. There are two kinds, shown in Figure 46. The old-fashioned ones are diamond-shaped pieces of flexible metal, sharp at both ends, and the easiest way to install them is with those needlenose pliers. There's another kind of point, however, which can be more easily and safely slipped into place with some pressure from a screwdriver blade. Seat the glass into its bead of glazing compound and fix it with the points.

The final step is putting the finish glazing compound in place. Take a glob about the size of a golf ball from the can and roll it between your hands into a long snake. Then, starting at the top, press the snake, or bead, of compound into place as illustrated (Figure 47). You can smooth it with your finger, or with a screwdriver blade, or with a putty knife, if you have one. You want to make a firm fill, so work your way down little by little instead of trying to put the whole snake in at once, and you'll run less chance of leaving air bubbles underneath.

Just as with caulking compound, you'll find that glazing compound gets fractious in very cold weather and won't want to cooperate with you. Let the can of compound stand indoors to get warm before you use it in the winter; keep it under your shirt next to your body when you go outdoors, and work that snake as fast as you can if there's a nip in the air.

If, after both the frame caulking and window glazing, you find you still have a problem, then weatherstripping is the next step.

Casement windows can be treated much like entry doors, with the bottom being treated like the top instead of like a sill, since you

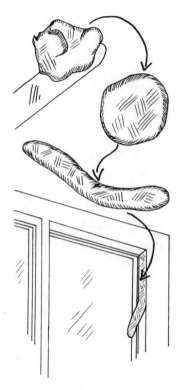

Figure 47 Glazing compound applied to seal a window pane.

don't have the wear factor to contend with as you do with a doorway.

Conventional windows—what carpenters and builders call double-hung sash—are trickier. As we mentioned earlier, windows that are never opened should be caulked around the cracks for a permanent seal. There are several options for weather-tightening windows that are openable.

In some double-hung windows, you'll be able to accomplish a closure with that thin spring metal we mentioned earlier. The stuff isn't too practical for most door installations, but it can work with windows, as shown in Figure 48, and it will wear well over the years.

Another option is the metal-reinforced plastic gasket. This isn't going to look too dandy, but it will turn the trick if you have a noticeable problem with leakage. Installation is relatively simple following the drawings in Figure 49.

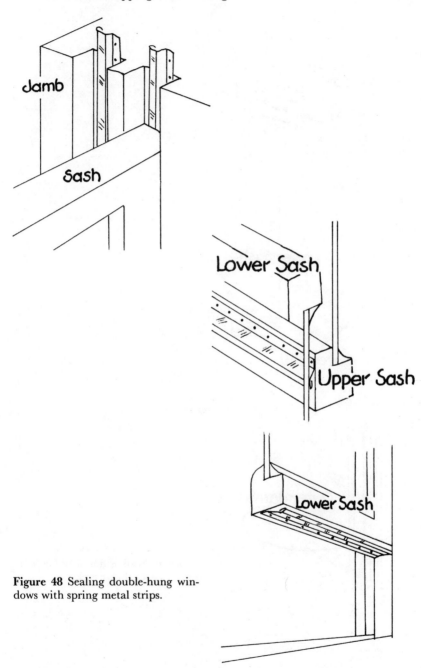

Figure 48 Sealing double-hung windows with spring metal strips.

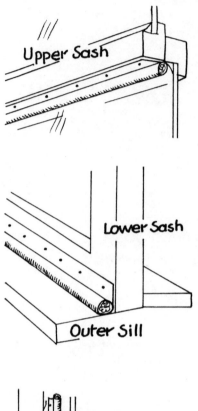
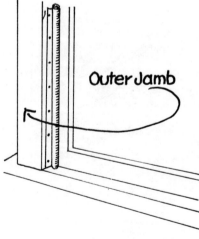

Figure 49 Sealing double-hung windows with a tubular gasket.

Weatherstripping and Caulking

Foam strips backed with adhesive can be used, within limitations. On the bottom of a window, for instance, where the foam will be compressed when the window is closed, foam will work satisfactorily. Up the side, where the foam will be subject to abrasion when the window is opened and closed, the stuff won't last very long.

There are so many different kinds and designs of windows that it isn't possible to deal with every situation here, but perhaps the principles outlined and the materials described will help you find your way to the solution for your specific problems.

IN SUM

Just to round things out before you go on to the next chapter, let's do a quick review of the basics.

Caulking, pointing, glazing and weatherstripping—in each case you have two ends in view: (1.) To keep the outdoors out and the indoors in; and (2.) To use techniques and materials of maximum durability so your work will last as long as possible.

In general, no expensive tools are required, and the working materials are inexpensive enough to more than justify the cost in relation to the savings that can be effected. If you do only part of the job, you'll still be saving money because every crack filled, every window fixed, each door weatherstripped contributes to the goal. In the long haul, doing part of the job, but doing it thoroughly, will be more advantageous and economical than taking a slapdash swipe at the whole thing so that no part is really done right.

Finally, a little nudge on that annual checklist. Riffle through this chapter again with paper and pencil and compile the roster of the things you should examine. When you get done, put that list in a place with the permanent records so you can refer to it every summer when the weather is right for working outdoors. These chores are better tackled a piece at a time rather than hastily over the Thanksgiving weekend with gloves on and a muffler around your neck.

CHAPTER IV

Properties of Insulating Materials

Larry Gay

Great effort has been expended on research and development of insulating materials since World War II. However, much of this work has been on far-out subjects such as the insulation of space capsules, and we still don't know for sure—to take one example—how well urea formaldehyde foam will perform in the walls of a house here on earth. Nor do we know enough about the long-term performance in humid climates of fire retardants applied to loose-fill cellulose. And how do you keep mice from tearing up fiberglass? There's a lot to learn.

To answer some of these mundane but important questions, the Department of Energy has initiated the crash research program referred to in Chapter I. Objectives, among others, are:

1. To develop standard test procedures for determining the R-value of thick layers of low-density insulation;
2. To look into smoldering combustion of insulants;
3. To investigate the behavior of fire retardants in moist environments;
4. To determine the effects of heat and moisture on insulants;
5. To see what health hazards are associated with various insulating materials;

6. To evaluate the performance of highly insulated buildings during a fire;
7. To assess the safety of electrical wiring in highly insulated buildings;
8. To devise computer models that accurately predict the thermal behavior of buildings of varying construction.

Even when this crash program is completed some of the answers will be only tentative. To know for sure how well an insulant performs in real houses lived in by real people will take decades. Insulant A may have a higher R-value than insulant B, but if it turns out to shorten the average life of houses, insulant B could be the better choice. Unfortunately long-term statistical data simply is not available, so we can't put it in our equations. The situation may not be like flying blind, but visibility is not unlimited, either.

A survey of the literature shows that some fairly straightforward bits of information, on which decisions should be based, are either confusing or contradictory. For instance, if one looks at manufacturers' specs for the upper working temperature of urea formaldehyde foam one finds a range of 120–320°F.[1]

Another problem one runs into is that the properties of all insulants vary to some extent with density, care with which they are installed, weather conditions, and details of the manufacturing process. Thus all tabulated R-values, flammability ratings, etc., are in a sense nominal. With this in mind, here is a necessarily incomplete rundown of the properties of the major building insulants currently in use in North America.

FIBERGLASS

More than 85 percent of all residential insulation in the United States today is fiberglass. This is a clear testimonial to the suitability of fiberglass as an insulant. It has a reasonably high R-value, does not readily absorb water, is nonflammable (although its resinous binder is), does not rot, and is fairly inexpensive. No wonder it is so popular. If fiberglass is to be criticized, it might be because its R-value per inch is low compared to some foams, and because there is considerable energy consumed in its manufacture.

A current process for manufacturing fiberglass insulation involves a first stage of melting the glass, followed by extrusion through a whirling hollow cylinder with peripheral perforations. Then steam jets impinge on these fibers and attenuate them to a diameter as low as .001 inch. The resulting wad of glass wool is passed through a spray of thermosetting resin to bind the fibers together and then is compressed to the desired density. Next it is baked to set the binder, and finally it is sliced into batts, blankets, or, just recently, rigid panels. There are several variations on this basic theme.

This process is energy intensive because of all the heat needed to make glass in the first place from quartz, limestone, and soda ash. In some cases molten glass is fed directly to the fiber-making apparatus, whereas in other cases marbles are made as an intermediate step. This requires a second melting before extrusion of the fibers. One may conclude that the price of fiberglass will be closely tied to the price of energy, and therefore will probably rise steeply with the cost of energy in the near future.

The R-value of fiberglass, like that of all air-trapping insulants, depends on its density and temperature. The relationship is graphed in Figure 50. As the density increases, the average size of

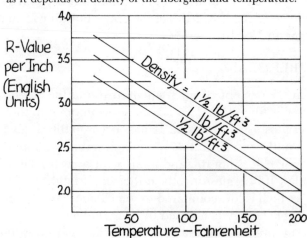

Figure 50 Graph shows the R-value of fiberglass insulation as it depends on density of the fiberglass and temperature.

trapped air pockets decreases. This tends to reduce small-scale convection currents within the material. The R-value is highest at a density of around 5 pounds per cubic foot. At higher densities conduction through the glass fibers becomes dominant and the R-value falls. Actually the R-value depends as well on the fineness of the fibers, with small-diameter fibers giving higher R-values at equal densities. Fibers that are very fine are unsuitable for insulation, however, because they easily crumble.

A nice feature of fiberglass is that it is very resilient, having little tendency to mat together unless it becomes water-logged. The practical consequence is that loose-fill fiberglass can be blown into walls with very little settling.

Because of both the physical and chemical similarity of fiberglass to asbestos (both are silicates), there is a suspicion in some quarters that breathing glass fibers may cause fibrosis and/or cancer of the lung. In 1976 the United States Consumer Product Safety Commission (CPSC) received a petition from the Consumer Office of the Denver District Attorney requesting that a mandatory standard be established to reduce the "risk of injury of cancer in the respiratory and gastrointestinal tracts due to the inhalation and ingestion of particles from fibrous glass home insulation, during installation and by spreading of particles after installation." After reviewing the available information the CPSC concluded that there was not enough evidence to warrant such a standard.[2] Animal studies are now in progress under the auspices of the National Institute of Environmental Health Sciences to assess the risk associated with breathing all kinds of mineral fibers, including asbestos, rock wool, and glass. The results should be available in 1981. Meanwhile manufacturers have agreed to label fiberglass products with a warning to protect eyes, lungs, and skin during installation. Use of the standard vapor barrier with fiberglass should keep fibers from becoming airborne, as they might otherwise do in some overhead installations.

Paper backing on fiberglass batts and blankets and the binder holding the fibers together are both flammable. If the paper is impregnated with asphalt, the flammability is greater. The fire hazard is comparatively minor, but not altogether negligible. It is one reason why fiberglass should be kept away from light fixtures and chimneys. Another reason to avoid insulating around heat sources is that circulation of air around them carries the heat away and prevents the temperature from climbing dangerously high.

OTHER MINERAL FIBERS

Rock wool is similar to fiberglass, except that it is made from rocks such as limestone and shale. Slag wool is made from byproduct slag produced in the manufacture of steel and other metals. What is called "rock wool" is in fact usually slag wool. Often the term "mineral wool" is applied to slag wool, although strictly speaking, mineral wool is a generic term covering this group of products as a whole. The properties of rock wool and slag wool are for practical purposes identical to those of fiberglass. In the form of loose-fill they can be blown into walls and attics.

Asbestos is the only natural mineral fiber known. It has been replaced by other insulants to the point that it is hard to find it in a new house today. The severe health threat associated with breathing it was well documented in the 1960s.

If you have a boiler in the basement, it and the pipes coming from it may be covered with a white insulating plaster usually referred to as "85 percent magnesia." Until recently the other 15 percent was asbestos, but new 85 percent magnesia is apt to contain fiberglass as the supplement instead. Elsewhere, asbestos may be found in old acoustical tiles or as sprayed-on thermal and acoustical insulation in schools and other public buildings. It is not used in building insulation today, although it may be encountered in odd items such as stove pads and hair driers. Asbestos cement board is sometimes recommended as a cover for plastic foam insulation.

CELLULOSE

Cellulose fiber is no doubt one of the oldest insulants. In a broad sense, all cotton clothing is cellulose insulation. Cotton waste from the clothing industry is in fact used as a building insulant, although not extensively. Eel grass, shredded coconut shells, and shredded sugar cane are also cellulosic insulants. As anyone who lives in a wooden house knows, the key to successful use of cellulose is keeping it dry. If high and dry, wood, which is cellulose and lignin, lasts indefinitely. In European castles one can see sound wooden beams barely supported by disintegrating masonry piers.

By far the most common form of cellulose building insulation today in the United States is loose-fill, which is made from recycled paper. The paper is shredded, then pulverized in hammer mills, and then treated with dry fire retardant, usually boric acid or ammonium sulfate. You get almost as much R for your money with cellulose as with fiberglass. However, with cellulose, the R-value you pay for and the R-value you get may be two different things. Seven of eight samples tested in a government laboratory had R-values 11–63 percent below manufacturers' advertised values.[3] As energy costs rise, cellulose should gain a price advantage. After all, the major costs of manufacture are paid for the first time around. In an ideal world, the environmental plusses of cellulose would be neatly wedded with nonflammability, low moisture absorption, resistance to rot, and unattractiveness to ants and mice. In addition, the fire retardants would not leach out or cause corrosion of electrical fixtures. However . . .

In walls cellulose may settle with time. This is not a disaster as long as access is provided and it can be topped up from time to time. It is easily applied in attics; treated with a fire retardant, it can receive a Class I fire rating, which is tops. Under no circumstances should cellulose be allowed to get wet, since this washes out the fire retardant and compresses the fibers into a mass not much different from papier-mâché. Whether cellulose insulation retains its insulating and fire-resisting properties in very humid climates over decades is an interesting but as yet unanswered question. If the cellulose comes in bags labeled as meeting Federal Specification HH-I-515-D, you know it has been tested and meets current standards for flammability, moisture absorption, and corrosiveness to metals.

POLYSTYRENE

This is a spinoff from the synthetic rubber industry. The material was developed very rapidly during World War II. GR-S, as it was called, stood for Government Rubber-Styrene. Building insulation is made from polymerized styrene either by extrusion or by molding of expanded polystyrene beads.

In the latter process, the starting material is small, fairly dense beads which contain butane, pentane, or some other volatile gas. When the beads are heated by steam the gas expands, puffing up the beads. Further steam heating in a mold fuses the beads into a large block of beadboard, a very light material with good R-value per dollar. The block is then cut into slabs by pushing it through an array of parallel hot wires.

Beadboard is not impermeable to water vapor and must not be used without a vapor barrier, as some, alas, have found out the hard way.[4] In thicknesses of less than 4 inches it can be cut with an ordinary handsaw. Because it comes in big (4 × 8-foot) sheets of any desired thickness, beadboard fits nicely into post-and-beam construction. Unfortunately it tends to warp a bit. This makes it less suitable for window shutters than extruded polystyrene or urethane.

Note that petroleum products are used in the manufacture of beadboard. Turning to the question of insulating with a petroleum product to save petroleum, even the most reckless calculation will show that the polystyrene used in insulation saves energy in an amount far greater than its own energy content—that is, the amount of energy that would be liberated if it were burned. Similarly, fiberglass saves much more energy as insulation than is needed for its manufacture.

Extruded polystyrene is more expensive than beadboard (see Table 5). To make extruded polystyrene, a volatile liquid such as methyl chloride is added to molten polystyrene under pressure. The melt is extruded through a large orifice, cooled, and then cut into boards. The density of the product is a bit higher than that of average beadboard, and has a higher R-value. The compressive strength of extruded polystyrene is also higher than that of beadboard, making it especially suitable for use on top of roofs or to face exterior basement walls.

Polystyrene is fairly hard to ignite but burns with a dense, black smoke; it should be covered with a thermal barrier to prevent ignition. Either form may or may not be treated with a fire retardant.

Styrofoam is a trade name for extruded polystyrene made by the

TABLE 5
Relative Prices of Major Insulants*

	R-Value/inch	¢/R unit/ft²
fiberglass batts or blankets	3.2	1.2-1.5
cellulose loose-fill	3.5	1.6-2.3
polystyrene beadboard	4.0	2.5-3.7
extruded polystyrene	5.0-5.5	5.0-6.0
polyurethane	6.3	6.0-7.1
vermiculite	2.0-2.5	5.0-5.5

* These figures are as of July 15, 1979 and only approximate, since price per inch depends on factors such as board thickness, batt thickness, and density. The R-value per inch is also somewhat variable. Nevertheless, the tabulated values give a good idea of relative prices of the major insulants.

Dow Chemical Company. It has come to be used so widely that many people, even some who sell insulating materials, refer to all polystyrene foams as Styrofoam.

POLYURETHANE

Like polystyrene, polyurethane (also called "urethane" for short) is a manmade polymer which can be blown up into a cellular structure of very low density. Its R-value is significantly higher because the blowing agent, a gas of low thermal conductivity, remains trapped in the foam. The identity of the gas varies, but a common one used in the United States is trichlorofluoromethane, a nonflammable refrigerating gas also known as Freon-11. Its thermal conductivity is about one-fourth that of air. Freon-11 is not very toxic to human beings, but is under suspicion as a member of a chemical family thought to reduce the concentration of ozone in the upper atmosphere—ozone which protects us all from getting too much ultraviolet light from the sun. It would be silly to ban the use of these chemicals as aerosol propellants but allow their use in insulating foams, since eventually they must escape. Nevertheless, this is the current situation. One hopes the industry will find a more benign blowing agent.

Polyurethane has a number of applications other than as a building insulant, and is produced in several different densities. The R-value of insulating polyurethane is often given as 7 or more, but it tends to decrease with time. The reason for this decline is an exchange of gases with the atmosphere and some shrinkage directly after manufacture. Within a year the R-value drops to about 6.3 where it remains indefinitely.

Polyurethane degrades rapidly when exposed to sunlight, so it should be covered if used outdoors—for example, on pipes connecting a solar collector to the house. It is an extremely good pipe insulator because its high R-value permits use of a relatively thin layer which keeps the surface area (and therefore heat loss) to a minimum. However, its maximum service temperature is around 200°F. Thus it is unsuitable for superheated steam pipes.

Polyurethane is difficult to ignite, but can burn in the presence of an ignition source of sufficient size. It burns with a good deal of smoke, and this smoke contains hydrogen cyanide gas, which is lethal. This is why we recommend using polyurethane insulation externally. Relevant codes require covering it with at least a ½-inch thickness of gypsum board securely fastened to the wall framework. The gypsum board is supposed to prevent the foam from catching fire for at least 15 minutes (enough time for the fire department to get there?) when the temperature on the room side of the wall is above 1000°F. Directly applying plywood or particle board over the foam is not considered adequate protection. When the foam is greater than four inches thick, sprinkler protection is also recommended.[5]

Polyisocyanurate is chemically and physically a very close relative of the better-known polyurethane; it is also puffed up with fluorocarbon blowing agents. Polyisocyanurate was first made in 1961, whereas polyurethane was developed in Germany before World War II. Considerable attention is now being paid to polyisocyanurates because some types of this polymer are highly flame resistant. The hope is to make a copolymer of isocyanurate and urethane with more flame resistance than urethane and less brittleness than isocyanurate. Another advantage of isocyanurate is that its upper working temperature is 300°F.

Because they are expensive, both polyurethane and polyisocyanurate are best limited to special applications where a thicker layer of a lower-R foam would not be as good, that is, they make most sense as pipe insulants and as window shutters.

UREA FORMALDEHYDE

This foam is comparatively cheap and has the outstanding advantage that it can be injected evenly into any wall cavity. Its disadvantages are that it may shrink after injection and it may give off lingering formaldehyde vapors.[6]

Unlike the other foams discussed so far, U-F foam is generated at the site by mixing the U-F resin with a catalyst and detergent, and then blowing the frothy product into a wall with compressed air. One or more holes must be drilled for each stud cavity and plugged at the end of the job or, if feasible, siding boards may be removed to permit access to the inside of the wall. The foam sets very quickly. Water vapor and formaldehyde are then gradually replaced by air in the open cell structure of the foam over a period of months.

Because quality control cannot be as good at the job site as it would be in a factory, the quality of U-F foam insulation varies tremendously from job to job and from operator to operator. Foams differ in chemical composition, depending on what additives are mixed with the resin at the factory. Some are inherently better than others. Even with a good brand of foam, a good job cannot always be counted on since the quality of the foam depends on other things such as: the temperature and humidity on injection day, the ratio of the ingredients and how they are mixed, how old the ingredients are, and the type of injection apparatus used. Some people who have had their houses foamed with urea formaldehyde have been bitterly disappointed. It is absolutely essential to carefully investigate the insulating contractor and his record of success before hiring him, and then to insist that the contractor sign a contract.

Several studies of U-F foam are now in progress, one at the National Bureau of Standards. In the preliminary stages a wide variability among U-F foams was established. Clearly some foams suffer severe shrinkage and some foams break down at temperatures and moisture levels only slightly above what might be encountered in a typical house.[7]

In the nine months since a friend in the next town installed U-F foam, it has shown a linear shrinkage of 4 percent. In a study of 44 houses in Portland, Oregon, average linear shrinkage was found to be about 8 percent, that is the equivalent of a 2-inch crack in a 24-inch stud cavity.[8] Another study found area shrinkages of 10 percent and an additional reduction of area coverage of 10 percent because of careless installation techniques.[9]

What is the effect of this shrinkage on the insulating value of a wall? Figure 51 shows how the calculated effective R-value of a nominal 2×4 stud wall decreases as the urea formaldehyde shrinks, and Figure 52 shows the same thing for a 2×6 stud wall. (Shrinkages are commonly reported either in terms of linear percent or area percent. To a good approximation, area percent = 2 × linear percent.)

A point to notice about both these graphs is that shrinkages less than those reported above cause reductions in the effective R-value of U-F foam below what one would get with either loose-fill cellulose or loose-fill mineral wool, or with fiberglass blankets. Therefore, if serious shrinkage occurs you would have been better off installing one of the cheaper insulants in the first place.

True, cellulose may settle, but it can be topped up if access is allowed for. Actually, from a thermal standpoint, shrinkage of U-F foam can often be worse than is suggested by Figures 51 and 52. The reason is that when the foam is used in an old wall without an

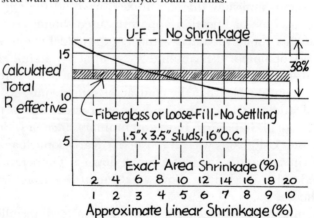

Figure 51 Calculated decrease in the effective R-value of a 2×4 stud wall as urea formaldehyde foam shrinks.

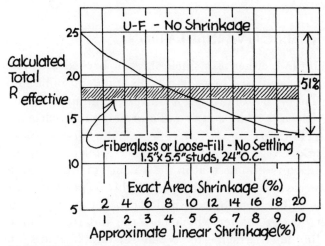

Figure 52 Calculated decrease in effective R-value of a 2×6 stud wall as urea formaldehyde shrinks.

adequate vapor barrier—often the case—slight infiltrating breezes can come through the cracks left between studs and receding foam. It should not be concluded that all brands of U-F foam always suffer excessive shrinkage, but surely it must be concluded that the homeowner should proceed cautiously.

Urea formaldehyde is generally not applied in attics because direct exposure to the atmosphere and high attic temperatures speed drying and cause excessive shrinkage and cracking. Furthermore, formaldehyde gas is heavier than air, and it settles into the rooms below. It is wise to use a vapor barrier with it. For one thing, its open cell structure allows water vapor to migrate through; for another, the vapor barrier keeps any formaldehyde fumes inside the wall. In almost all retrofit cases the most convenient vapor barrier will be a low-perm paint or impermeable wallpaper applied to the interior wall.

U-F foam will propagate a flame as long as an ignition source is applied, but cannot support a flame by itself. Its behavior with respect to fire hasn't yet been fully characterized, however, and the National Fire Protection Association has therefore taken the cautious attitude of recommending use of a thermal barrier with U-F foam—that is, $\frac{1}{2}$ inch of gypsum board over it.

REFLECTIVE INSULATION

Aluminum foil, either as a single sheet or backed with paper, has occasionally been used to insulate walls and floors of buildings. Today it is largely out of favor, not because it cannot be effective, but because it ceases to be a poor radiator and good reflector if it becomes wet or even damp.[10] To be effective, the foil must have a shiny surface and be adjacent to an air space, preferably $\frac{3}{4}$- to 1-inch wide. One can estimate that in the absence of aluminum foil heat transfer across such a vertical gap is roughly 60 percent by radiation and about 40 percent by conduction-convection. If one surface is shiny—it doesn't matter which one—the radiation component is nearly eliminated, which is to say that the total heat loss rate across the cavity is halved. The foil is even more effective if used in the floor where convection is relatively unimportant (because the warmer air is above the cooler air) and radiation correspondingly more important. If aluminum foil is used in a cavity, only one surface need be lined. Lining the second surface leads to only a slight improvement in R-value.

There are three other good reasons why aluminum foil is not more widely used. One is that several layers of it in a wall can act as a Dagwood sandwich of vapor barriers which trap moisture in the wall. The second is that mice and squirrels love to tear it up for their nests. And the third is that you get more R per dollar with fibrous insulants.[11] Like fiberglass, aluminum is energy-intensive and its price will be very dependent on the cost of energy.

Even so, there are situations where shiny aluminum foil is the insulant of choice. One place where it makes sense is between a steam or hot water radiator and the adjacent wall. Here there is generally not enough room for fiberglass or some other bulky insulant, but a single layer of foil reduces heat loss through that part of the wall directly behind the radiator by around 55 percent in the typical case.[12] Similarly, a wall may be protected from the radiant energy of a wood stove by use of a protective layer of shiny aluminum. It should not be paperbacked and should be positioned an inch away from the wall so that it may be cooled by air currents on both sides.

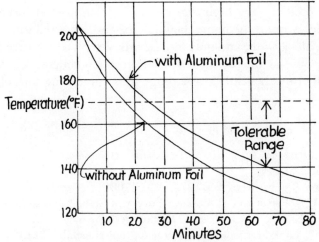

Figure 53 Coffee-cozy insulating dynamics: curves show the effect of aluminum foil on how fast the coffee pot cools.

One novel use of aluminum foil is as a coffee cozy. Figure 53 shows cooling curves for a cylindrical glass coffee pot commonly used in France and now in this country. If you consider coffee tolerable at temperatures between 170 and 140°F, a single layer of aluminum foil wrapped tightly around the pot with shiny side out extends the drinkable interval from 45 minutes to 60, and thus lets you read the Sunday *Times* in greater leisure. A good foam cozy would do even better.

VERMICULITE AND PERLITE

Vermiculite is expanded mica. Chemically it is a layered silicate rock containing about 12 percent water by weight. The rock is crushed and then quickly heated to about 1600°F. This causes the water to expand rapidly into steam and force the layers apart into accordion-like, flaky granules.

Perlite is made in nearly the same way. The difference is that the raw material is volcanic rock with a different crystal structure. The finished product is in the form of puffy white granules.

Both of these insulants are completely noncombustible, but of relatively low R-value. Their great advantage is the fact that they can be used at very high temperatures, in industrial furnaces for example. Around the home they may be easily poured into voids in cement block walls, and may also be mixed with concrete to raise its R-value, but at the expense of strength. Such concrete could be of use in footings beneath heat storage bins containing tons of rocks.

NOTES TO CHAPTER IV

1. W. J. Rossiter, Jr., et al., *National Bureau of Standards Technical Note 946,* 1977.
2. *Federal Register,* March 5, 1979, page 12081.
3. R. W. Anderson and P. Wilkes, "Survey of Cellulosic Insulation Materials," Energy Research and Development Agency, 1977.
4. This is one of those broad statements that is usually, but not always, true. Beadboard can be made in densities up to $3\frac{1}{2}$ pounds per cubic foot. Very dense beadboard has a permeability rating of less than 1, which qualifies it as a vapor barrier by former standards. Play it safe and don't use beadboard without a separate vapor barrier.
5. J. A. Stahl, "Using Foam Plastic Insulation Safely." *Fire Journal,* September 1978, page 43.
6. Fumes from urea formaldehyde insulation can be more than a nuisance. As this book goes to press late in 1979 the Massachusetts Department of Public Health has banned the use of U-F foam insulation in the state. A news report of the Public Health Department's statement says the department found "significant correlation between the UFF (urea foam) insulation and certain formaldehyde-linked illnesses, such as respiratory difficulties, skin and eye irritations, headaches and vomiting."
7. See Rossiter, Note 1.
8. George Tsongas, private communication. Department of Engineering and Applied Science, Portland State University, Portland, Oregon.
9. E. R. Vinieratos and J. D. Verschoor, "Influence of Insulation Deficiencies on Heat Loss in Walls and Ceilings." Paper delivered at the Department of Energy/American Society for Testing and Materials Thermal Insulation Conference, October 22-25, 1978, Tampa, Florida.
10. I. E. Smith and S. D. Probert, in *Applied Energy,* January 1979, page 85.
11. See Smith, Note 9.
12. See Smith, Note 9.

CHAPTER V

Retrofitting Insulation
Dana Zak

Retrofit is a new word, but American homeowners are learning its meaning quickly, of necessity. Retrofitting insulation is installing insulation in an existing structure. In this chapter we discuss important techniques in retrofitting insulation in different locations, using different materials.

MATERIALS HANDLING

When working with any type of wool batt or blanket insulation, it is important that you take precautions. You must take care to avoid inhaling stray particles or allowing them to be rubbed into your skin. For this reason it is recommended that you wear the following when working with fiberglass or mineral wool:

Wrap-around safety goggles;
Lightweight work gloves;
Respirator or cotton surgical mask (available at drugstores and most hardware stores);
Loose-fitting clothing (some contact with stray material is unavoidable, but wearing loose clothing helps keep wool insulants from being rubbed into the skin).

After working with insulants it is important that you take a *cool*, not warm, shower to wash away any material left on the skin. A cool shower keeps pores closed, and reduces the chance of skin irritation.

Cutting is the most time-consuming and exacting part of installing batt or blanket insulation. Therefore, everything should be done to minimize cutting and to make necessary cuts as accurate as possible. Wood framing in the home plays the largest role in determining whether or not to use batt or blanket insulants, and how much they will have to be cut.

Most existing homes have parallel framing members (studs, joists, rafters) spaced 16 or 24 inches on center (o.c.); and perpendicular members (firestops, bridging) 4 or 8 feet o.c., as in Figure 54. If the home to be insulated does not have the great majority of its framing 16–24 inches o.c., then batt or blanket insulation is not a good choice, because it comes pre-cut to fit these widths; using it in such a situation would require too much cutting to be practical. (It is less annoying to cut an equal amount of foam board or to use loose-fill.) Likewise, if bridging or firestops are generally spaced 4 or 8 feet o.c. then using batts instead of blankets is called for, because they are pre-cut to fit these lengths. A special situation is an attic which has a lot of cross-bridging between the joists (Figure 55). To fit batt/blanket insulation in and around cross-bridging requires extra cutting and handling of the insulation; having to deal with a lot of cross-bridging suggests the use of a loose-fill insulant.

The cutting of batt/blanket insulation that does need to be done can be accomplished with a pair of large shears for insulation that is

Figure 54 Roof, floor and wall framing. Fiberglass batts and blankets come precut to fit standard framing distances.

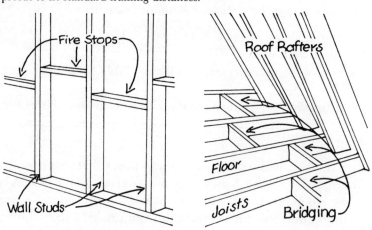

Retrofitting Insulation

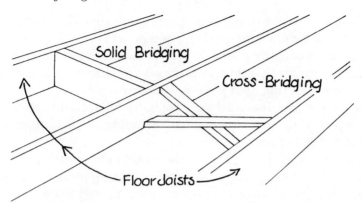

Figure 55 Cross-bridging makes laying insulation blankets between floor joists difficult.

3½ inches or less in thickness, or with a utility knife. Contractors invariably use a utility knife; a little practice with a sharp knife will enable you to cut all thicknesses faster than you can with shears. For cutting accuracy, a straight edge, which can be a level or a piece of framing, should be used to make cuts until you get an eye for cutting.

When cutting to size, allow a full extra inch of width and/or length. This ensures a good snug fit even if the cut is not the most accurate. If you are using batt/blanket insulation with an attached vapor barrier, and must cut it to fit a narrower width, you can form a new stapling flange by first cutting the vapor barrier and wool 3 inches over the width to be insulated, then removing 2 inches of wool, but not vapor barrier, from the cut edge. Figure 56 shows how the extra vapor barrier is then used for a new stapling flange.

Figure 56 Leave a stapling flange when making a batt or blanket narrower.

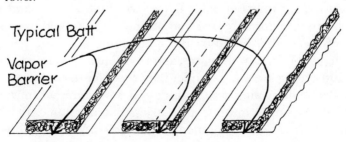

If you have a large number of spaces to fill, all of the same length, you can get the job done quicker if you pre-cut enough pieces to fill them instead of cutting and filling one piece and space at a time. Be absolutely certain that the spaces are really all the same length, and that you are cutting each piece to the correct size.

As you cut, put discarded pieces of batts/blankets into a garbage bag or pail. You can use them later for stuffing odd nooks and crannies at the end of the job.

Loose-fill insulation requires the same safety equipment and clothing as described above. If there is anything peculiar about its handling, it is the necessity to fluff hand-poured wools. Because they are compressed in their packaged state, pouring wools need to be expanded, or fluffed up to achieve maximum insulative value. The fluffing can be done by filling a cardboard box, or a large trash bag a third full with pouring wool, and shaking until the insulation fills the entire container. An unsatisfactory alternative is to pour the wool between attic joists, and then fluff it up with a garden rake. Don't bother.

Foam insulation boards are very easy to handle because of their light weight, rigidity, and inertness under normal conditions. Most people take no special precautions while cutting or installing them. Cutting boards can be done a number of ways, depending on the thickness of the insulation. Boards up to $1\frac{1}{2}$ inches thick you can either cut with a utility knife or score with a knife and then snap off. For thicker boards, a saw of some type is necessary. Circular, sabre, or jigsaws are fine, as are various handsaws. Accurate cuts are made by following straight edges or chalk lines.

ATTIC SPACES

In most cases, owners of older homes will find it easier and more cost-effective to insulate attic or ceiling areas rather than other parts of the house, because attic and ceiling areas are generally more accessible than wall spaces, and techniques used for insulating attic and ceiling areas are simpler than those for insulating existing walls. Before discussing actual insulation procedures we must warn those who intend to insulate attic spaces in older homes that have electrical wiring between or along ceiling joists. Where such wiring

Retrofitting Insulation

is subject to current overloading, placing insulation around it may be a fire hazard. The following was published by the United States Department of Energy in November 1978:

> When parallel nonmetallic sheathed cables [e.g., Romex brand] carrying 135 percent of rated current were placed between two layers of R-11 insulation, NBS [National Bureau of Standards] found that temperatures were more than double the allowed limit. The tests were considered conservative since they were conducted with new wires over relatively short runs. In addition, current rated at 135 percent may not be as common as 150 percent in those situations where over-fusing occurs. Overheated wiring circuits may ignite electrical insulation or adjacent building materials, or they may lead to gradual deterioration of the electrical insulation which could result in ignition from arcing or short-circuits.[1]

Further tests will be run in existing homes to determine whether results found in the laboratory hold true in the real world. Until those results are in, we feel obliged to urge homeowners to be cautious and not encase wires in thermal insulation.

If leaving your attic partially uninsulated seems to be a waste of heating dollars, keep these facts in mind:

1. The areas involved probably represent less than 5 percent of the attic;
2. Insulation can easily be added later, if doing so is proved safe;
3. Heat losses incurred are insignificant compared to the loss of a home due to an electrical fire.

A careful inspection of the area to be insulated is the all-important first step in insulating the older home. This is true not only for attic spaces, but for walls and floors as well.

Precautions in the Inspection. Be certain to take the following precautions when you are about to inspect an attic.

Provide adequate lighting. For safety's sake, take a flashlight when you make the inspection. For actually insulating, you'll need something better.

Provide something to stand on. If your attic doesn't have a floor, provide your own by using a piece of plywood, or several planks

(2×8's or larger). *Under no circumstances should you step between the joists, or on an unsupported end of temporary flooring.*

Be careful around rafters and nails. Don't bump your head into nails or rafters, especially those nasty roofing nails that seem to be waiting to puncture the unwary. The cautious person will wear a hard hat.

Do not touch exposed wiring—enough said.

Try to inspect on a cool, cloudy day. This advice holds true for insulating as well. Neither is a very comfortable activity; why make them worse by working in an oven?

Things to Look for. While making the inspection watch for the following faults.

Leaks in the roof. Dripping water is the most obvious evidence of a leaky roof. You should also look for discoloration of roofing and rafters, and any gross evidence of rot. It is a serious mistake to try to insulate before leaks are fixed. Water is your Number One enemy, here and elsewhere. Don't fool yourself; leaks and water damage will not repair themselves. Given time, they will worsen and cost you more money. Beyond this, if insulation is installed without repairing leaks, water can become trapped by the insulation and actually accelerate structural damage. One more time—fix those leaks, and remove any old insulation that has become water-logged.

Bad Wiring. If electrical wiring runs through the attic, look at it to see if there are any signs of deterioration: exposed wires, frayed electrical insulation, and broken insulators. If you don't feel competent to make this inspection, ask your local building inspector or fire department to do it. Remember, wiring problems, like a leaky roof, will not go away; they must be repaired.

Matted older insulation. Older insulation that has become matted or water-logged has little insulative value, and must be replaced. If insulation is matted or missing, it could mean that you have squirrels, mice, or other critters living with you. They love to make warm nests from insulation, be it new or old.

Miscellaneous. Locate any recessed light fixtures, ventilating fans, heating elements, and other heat-producing devices in the area between the joists. These items will require special attention; you'll

Retrofitting Insulation 105

have to keep insulation and vapor barriers at least 3 inches from them to prevent overheating and ignition. More on this later.

Where to Install Insulation. After making any necessary repairs, a decision needs to be made as to where to install the insulation in the attic. If the attic is unfinished, and you don't plan to finish it later, then insulating between the floor joists will save time and money. The only exception is if the attic has a nicely finished floor that might be difficult to take up. If the attic is finished, or if it will be finished later, then plan to insulate between the rafters and between the studs of the knee walls and end walls.

In attics that don't have floors and have little cross-bridging, if the joists are spaced 16 or 24 inches o.c., blanket insulation has distinct advantages. It is available with an attached vapor barrier,* which saves the step of installing a separate one. It is easy to maintain adequate clearances from heat-producing fixtures by simply cutting it to length. Finally, it can be installed by one person; blow-in requires two.

INSULATING UNFINISHED ATTICS WITH BLANKET INSULATION

Begin all work at the far corners of the attic and work back to the access hole. If there is no old insulation, use blankets with an attached vapor barrier. If there is old insulation, you can cover it with unfaced blankets, and paint the ceiling of the room below with a low-perm paint which will form a vapor barrier (Glidden's Insul-aid, for example). When using blankets with attached vapor barriers, place the kraft or foil barrier *down* against the ceiling below. Push down the insulation carefully until it comes into contact with the ceiling. Don't try to staple it into place.

Remember: do not encase electrical wiring in the insulation and do not place the insulation closer than 3 inches away from recessed lights, ventilation fans, radiant heat fixtures, chimneys, etc. Some manuals suggest that it is possible to insulate over some electrical

*See page 50 for a discussion of the advisability of having a vapor barrier in the attic floor.

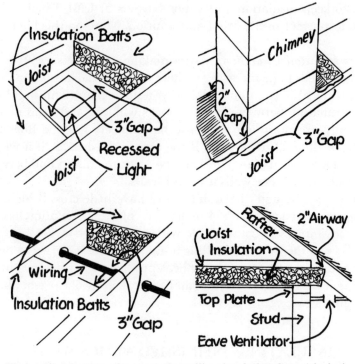

Figure 57 Adequate clearances for ceiling insulation from (clockwise from upper left): light fixtures, chimney, roof boards, wiring.

fixtures if an airway is maintained. We strongly urge that you *do not*. For those very energy-conscious, there is only one way to handle this problem: remove these fixtures and plug up the holes.

Maintain a ventilating airway at the ends of joists so that air entering ventilators under the eaves is free to flow along the rafters and the underside of the roof. Insulate over the top plate as far as possible without blocking the airflow.

If, while using blankets, you run into much cross-bridging, you'll understand quickly why some people prefer to use loose-fill. Cutting the blankets to fit in and around cross-bridging is a nuisance. After cutting a length of blanket so that it reaches the middle of the cross-bridging, you will need to cut a notch about 4 inches long in the center of the end to fit into the bridging. The insulation is then woven into the cross-bridging so that one end goes under, the other

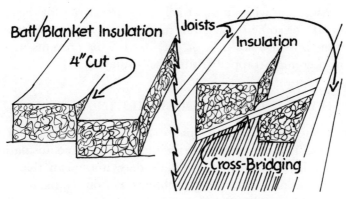

Figure 58 Weaving blanket insulation around cross-bridging.

over one side, as shown in Figure 58. The process is repeated on the other side.

After all insulation is in place between joists, the access hole cover can be insulated by stapling or gluing a piece of blanket to it with a waterproof adhesive.

If you intend to add a second layer of insulation, there are three things to remember:

1. Observe all clearances listed above;
2. Run rows of the second layer of insulation at right angles to those of the first. This blocks thermal bridges formed by joists (Figure 59);

Figure 59 In a two-layer insulation job the second layer should be at right angles to the first.

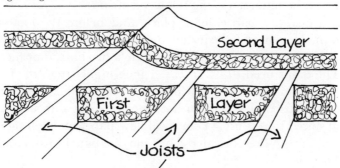

3. Use insulation without an attached vapor barrier for the second layer. If, for some reason, you must use some with an attached barrier, punch holes in it or peel it off. This prevents moisture from becoming trapped within the insulation.

HAND-POURING LOOSE-FILL IN ATTICS

In attics where joists are not spaced 16 inches or 24 inches o.c., or where there is a lot of cross-bridging, hand-poured insulation is easier to install than blankets. This does not mean that hand-poured insulation is without its problems. For one thing, it does not form its own vapor barrier.

There are two ways of applying a suitable vapor barrier for loose-fill. One is to paint the ceiling below with a low-perm paint. The other way is to cut 4- to 6-mil polyethylene film into strips to fit between the joists. Pre-cut strips *might* be available from your insulation supplier. If they aren't, the easiest way to make your own is to unroll a portion of folded-up film and cut it while it is still folded. Using a utility knife and a straight edge, you should cut the film 8 inches wider than the space between the joists. This oversizing allows enough extra film to make attaching it to the joists quite easy. If the attic is cramped, cut the film somewhere else.

In the attic, unfold the strips into the joist spaces, *being careful to keep the film away from lights and other heat sources.* If the strips aren't long enough for the spaces, overlap them at least 6 inches and tape the joints with duct tape. For holding the film in place there are two approaches, shown in Figure 60: stapling it to the joists every 12 inches; and using wood strips as shown. Of the two methods, a better seal can be obtained using the wood strips.

After the vapor barrier is in place, and before the loose-fill is poured in, dams need to be installed to prevent insulation from falling into those areas where it could create a fire hazard or block air flow. The dams need to be at least 1 inch higher than the level of insulation. You can make your own from sheet metal. Thin sheets of aluminum used in offset printing are easy to cut with shears, and are well suited to this use. (Your local newspaper may sell them for next to nothing.) Prefabricated dams may be bought from an insulation supplier.

Retrofitting Insulation

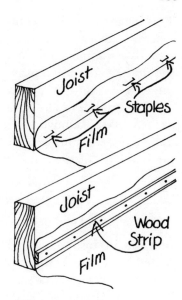

Figure 60 Wood strips nailed down over the vapor barrier film make a better seal than staples alone.

Once all dams are in place, insulation can be poured. Start as far as possible from the access hole and work back to it. Perlite and vermiculite can be poured directly from their bags, but pouring wools must be fluffed up before they are poured—as described in the materials handling section.

Some caution should be exercised when using vermiculite and perlite to avoid overloading the ceiling beneath. The U.S. Department of Energy recommends the following load limits, assuming a gypsum board ceiling:

Ceiling Thickness	Framing	Maximum Loading
½"	24" o.c.	1.3 lbs/sq. ft.
½"	16" o.c.	2.2 lbs/sq. ft.
⅝"	24" o.c.	2.2 lbs/sq. ft.

Because density varies from brand to brand, it is impossible to give a maximum allowable depth for every ceiling thickness and framing center combination. But, because each bag of insulation is labeled with weight and typical area of coverage at a given depth, it is easy to determine loading by doing a little arithmetic.

For example, perlite can weigh 2 to 11 pounds per cubic foot.

110   THE COMPLETE BOOK OF INSULATING

This means that the lightest types weigh only 1 pound per square foot at a depth of 6 inches. The heaviest forms weigh almost 1 pound per square foot at a depth of *1 inch*. It is clear that care should be taken when using perlite to achieve high R-values in the attic.

After insulation has been poured, it can be leveled with a light garden rake. Remove any insulation that has fallen outside the dammed areas. Finish insulating by covering the access door with a piece of batt/blanket insulation. Staple or glue it in place.

BLOWING IN LOOSE-FILL INSULATION IN ATTICS

Blowing in insulation with a rented blowing machine is often touted as the quickest, simplest way for homeowners to insulate attic spaces. For some areas, such as in cramped spaces and under floors, this *may* be the case, but a number of factors suggest that blowing in insulation should be done only as a last resort:

It requires two workers—one to spread insulation, the other to feed insulation into the machine;

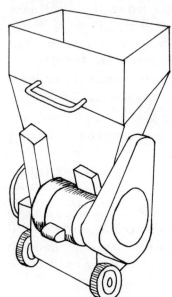

Figure 61 Insulation blower for loose-fill.

The worker is dependent on the condition of the machine (gas engines may be difficult to start and electric motors may blow fuses);

Special safety precautions must be taken when using the blowing machine. Many particles will be let loose into the air, creating an increased health hazard;

Because it is so easy to shoot insulation into cramped areas and under floors, there may be a tendency to ignore doing the work necessary to maintain adequate clearances around heat-producing fixtures—a potentially fatal mistake.

If you must blow in insulation, observe the following precautions:

Do not allow children near the machine while it is in operation.

Do not put hands into the hopper to remove clogs, even if the motor is not running. It is possible that tension could remain in the drive mechanism and removing a clog could allow the blades to spin. Use a tool, not your hands.

Keep everything but insulation from falling into the machine. One way to do this is to place only insulation on the loading ramp.

When you're ready to insulate, place the machine as close as possible to the attic access door and run the flexible hose up to the attic. If you don't have a machine with a remote control on/off switch, take time to get Stop and Go signals straight with your helper.

Before feeding insulation to the blowing machine, the helper should allow it to run for half a minute to let it get up to speed. Then, rather than dumping a whole bag into the hopper in one lump, the helper should feed in the insulation, being careful to break up any tightly compressed wads before they go into the hopper.

Actual installation in attics without floors is much the same as with hand-pouring, including raking to level the insulant. Several depth indicators, made of scrap wood and attached to various joists can help get the right amount of insulation applied.

The first step in insulating beneath attic floors is to locate any heat-producing devices below the boards. To some extent this can be done by looking at the ceiling of the room below. In any case, you must take up some attic floorboards in order to place dams at rafter ends. You'll need to take up others in the center of the floor so that it is possible to check for bridging which might block the flow of insulation.

Once you have a number of boards removed you should be able to see with the aid of a mirror and flashlight whether there are any fixtures or wiring that require installation of dams. Next, you'll have to pull up enough boards to install dams. This could involve considerable work in some attics.

Finally, loose-fill is blown into the joist spaces through gaps left by removing floorboards. The vapor barrier, if any, must be in the form of paint on the ceiling of the room below.

Now, to see whether you have a done a good enough job, some still, cold morning before the sun is up, carry a thermometer up to the attic and measure the temperature. If it is more than a few degrees higher than the outside temperature, there is a heat leak into the attic somewhere. This is most apt to be a convection current coming up around a chimney, light fixture or vent. Find it and take care of it.

INSULATING FINISHED ATTICS THAT HAVE KNEE WALLS

Insulating attics with knee walls is easy, if there are access doors to the spaces behind the knee walls (see Figure 62). If there aren't any, you'll have to cut into the walls and make doors of your own.

To start, you must determine if there is any electrical wiring running along the knee wall. If there are no outlets or fixtures, then you can feel relatively certain that you won't run into any wiring as you cut into the wall. If there are outlets, then you should assume that there is wiring behind the wall. To be safe, kill the circuit before cutting an access hole above the level of the wiring. If there is any overhead lighting, pick a joist space as far as possible from the fixture(s) for cutting into the wall.

Of course, it is necessary to find two studs before cutting between them. Sometimes they can be located visually if the job of finishing the wall was done poorly, or not at all. Rows of nail heads or holes will mark the location of studs.

If this lazy way of finding framing members is not possible, then

Retrofitting Insulation

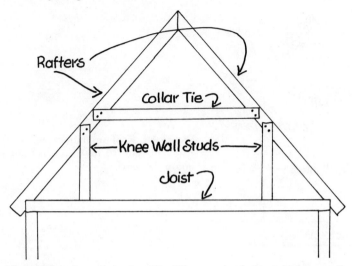

Figure 62 Attic construction. You'll have to get behind the knee walls to insulate them.

try either of these: 1. Tapping the wall with a hammer while listening for the less hollow, more solid sound of the framing; or 2. Drilling a hole and using a stiff wire to probe for framing.

After gaining access to the area behind the knee wall, start insulating the portion of the ceiling that angles up from the knee wall to the collar ties. The only way to do this effectively is to push batts up from below (Figure 63). You can do this if you have enough room beneath the roof to maintain a 2- to 3-inch ventilating airway, and the length to be insulated isn't much more than 3 to 4 feet.

Use a stick to push the batt up, and maintain the airway. Be careful not to compress the insulation too much. During this operation there is a good chance of tearing an attached vapor barrier.

The next area to insulate is the knee wall. Insulation of choice here is batt/blanket with *reverse* stapling flanges. These allow you to place the attached vapor barrier next to the wall and to staple on the side facing you, as in Figure 64. Unfortunately, insulation with reverse flanges is not easily obtained everywhere.

If reverse flange insulation is not available, use clips cut from coat hangers or stiff wire to hold unfaced blanket insulation in place. The interior surface of the wall may then be covered with a vapor-

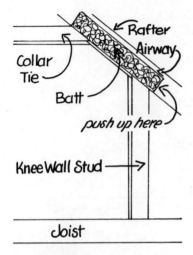

Figure 63 Roof insulation with fiberglass batts. Be sure to leave a 2-inch air space above the batt.

barrier-forming paint or a vinyl wallpaper, if you did not use insulation with an attached vapor barrier.

After wall insulation is in place, install insulation between the joists as you would in any attic space, following the same precautions and procedures, including maintaining adequate clearances. This finishes insulating behind the knee wall. The next step is to insulate above the collar ties as you would in any other unfinished attic area.

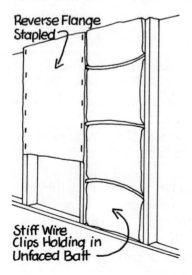

Figure 64 Two ways to hold knee wall insulation in place.

To cover the access holes you've cut into the walls and ceiling, build simple covers from plywood. They must be weatherstripped well, or air infiltration will negate all the insulating that has been done. Install the door by *screwing,* not nailing it to the framing. Finish as you would any other portion of the wall.

CATHEDRAL CEILINGS, FLAT ROOFS

There are two good ways to handle cathedral ceilings. One is to reduce the volume of the room by building a drop ceiling. The other is to install foam boards on top of the roof outside the building. Putting foam or fiberglass directly under the roof has been done, but there are several disadvantages to this. One is that it is inevitably extremely awkward and inefficient work. Another is that there must be an extra good vapor barrier because there is usually no ventilation between insulation and roof. Fiberglass cannot be used at all unless the rafters are spaced at standard distances, that is, 16 or 24 inches o.c. If, in spite of this advice, you go ahead and install foam boards below the ceiling, you should cover them with gypsum board as a fire barrier, and that means extra weight which the roof may not be designed for. Fiberglass may be covered with something lighter, such as thin paneling. But with either gypsum board or paneling you will need an extra man and staging to do the work unless you are over ten feet tall.

A more satisfactory solution is shown in Figure 65. Lay extruded polystyrene, urethane, or isocyanurate boards on the existing roof, outside, and then cover them with plywood and asphalt roofing of one kind or another. This way you are working downward, high R-values can be obtained easily, and the old roof makes an excellent vapor barrier. Another important advantage is that the rafters do not act as thermal bridges as they would in the case of fiberglass below the roof. If you do the work when new roofing is required anyway, the cost of part of the job can be charged to general maintenance and not to insulating. With a flat built-up roof, foam boards also work well. Tapered boards installed by a contractor insure that water runs off in the right direction.

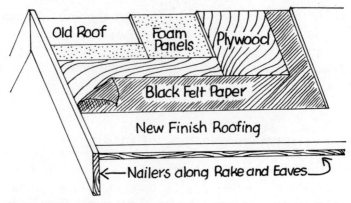

Figure 65 Insulating above an existing roof is easier and more effective than insulating the roof from below.

INSULATING EXISTING WALLS

Insulating existing walls in older homes is no weekend project; there is little, if anything, the average homeowner can do without the aid of a contractor. The approaches to insulating walls are:

1. Blowing wool, cellulose, or U-F foam into the wall;
2. Tearing apart the wall and adding insulation (see Chapter VI);
3. Building a new wall inward.

We cannot recommend adding foam board insulation to the exterior of existing walls, because of potentially serious moisture retention and damage. Old walls seldom have a decent vapor barrier on the inside.

There is no completely satisfactory insulation that can be blown into walls. Each of the three listed below has liabilities.

Blowing wools (slag wool, rock wool, fiberglass). These materials can get hung up on internal obstructions such as nails and electrical fixtures, and they may settle somewhat after installation. This means less than complete coverage. On the plus side, they are completely fireproof.

Blowing cellulose. When blown into walls, cellulose has an R-value a little higher than wools, and lower than the nominal value for urea formaldehyde. Cost is roughly equivalent with wools, and substantially less than U-F foam. Like wools, cellulose may get hung up on internal obstructions, but not to the same extent. Cellulose must be used with a good vapor barrier to prevent its becoming waterlogged, a situation which can cause a serious loss of R-value and perhaps leaching out of the fire retardant. We feel using cellulose in walls is chancy.

Blowing U-F Foam. A risky choice. (See Chapter IV.)

You should be aware that in order to blow any form of insulation into existing walls you must drill a large number (at least 2 per stud cavity per floor) of holes or 2 inches in diameter. These holes must later be filled with wooden plugs, then caulked, sanded, and painted. The plugs can be finished quite nicely, making them difficult to detect. Before insulating is done, there should be a clear understanding of whether the homeowner or the contractor will do the finishing.

INSULATING FLOORS ABOVE UNHEATED AREAS

Two choices are available here, depending on whether or not flooring will be replaced. If the house is to get a new floor, then the insulant of choice is *extruded* polystyrene used as subflooring (Figure 66). It is very resistant to crushing and in many cases will be a sufficient vapor barrier in itself. However, it is easy to lay a sheet of polyethylene over the insulation before the floorboards go down, so you might as well put down the poly as a vapor barrier.

If there is not going to be new flooring then reverse flange batt or blanket insulation with attached vapor barrier should be installed between the floor joists from below (Figure 67). It is supported from below by a layer of kraft paper, and its compressibility allows a good, snug fit. Unfortunately it is not always available.

Begin insulating by tucking the end of a batt or a blanket up against the header as shown in Figure 68. Staple it into place every 12 inches. The flange can be used on the inside or bottom surfaces of the joists. Smooth out the insulation as you staple so that it doesn't bunch up.

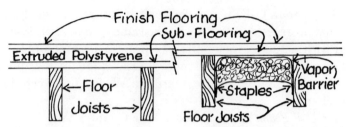

Figures 66 and 67 Two approaches to insulating a floor: foam boards above the joists (66) and fiberglass between them (67).

If there is any cross-bridging between the joists, you're unlucky. You'll have to make a notch down the center of the insulation and weave it into one side of the cross-bridging, then staple as you normally would. The process is then repeated on the other side.

Some crawlspaces are so shallow that there is virtually no room to wriggle into them, let alone insulate them. There is little the homeowner can do to insulate such areas except to refloor as described at the beginning of this section. If the owner cannot or will not do this, the only recourse is to tighten up these areas with weatherstripping and caulking to reduce air infiltration. Then put down a thick rug. Once a crawlspace has been made tight, it is very likely that ventilation in summer will be needed. If moisture in summer is excessive, a poly vapor barrier can be laid directly on the ground.

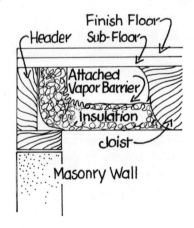

Figure 68 Floor cross-section with joist cut away to show how the insulation blankets are laid in. Be sure to insulate the header, too.

INSULATING HEATED CRAWLSPACES AND BASEMENTS

Insulation may be applied externally or internally to these structures. Of the two approaches, external application is better, because it adds the masonry walls of the crawlspace or basement to the heat-storage capacity of the house, thus reducing temperature fluctuations. Another advantage of external insulation is that, with it, flammable materials are kept out of living spaces.

The major problem with an external retrofit of basement and crawlspace walls is that these portions of the house below grade must be excavated. However, you need not dig all the way to the base of the foundation. Figure 69 shows a scheme that provides good insulation without requiring a ditch digger. The horizontal foam board is almost as effective as if it were laid vertically against the wall. This follows from the fact that heat flow near the foundation in winter tends to be up toward the coldest temperature. Just as water runs from high to low, heat runs from hot to cold. Thus a horizontal board below grade can be a good heat dam.

Internal Insulation of Crawlspaces. If your house has a crawlspace used to conduct hot air you can apply insulation internally by the following method. Lay a 6-mil polyethylene vapor barrier across the entire floor to prevent drawing excessive amounts of moisture from the earth. Don't put the entire barrier down at once, or you will be walking all over it as you work. The film should first be laid along one of the walls where floor joists butt into the header joist at right angles. You can use duct tape to hold the film in place (Figure 70). Repair any tears in the film with duct tape.

Cut blanket insulation with attached vapor barrier so that the entire piece is 2 feet longer than the distance (X in Figure 71) from the top of the header to the floor below. Cut enough pieces of insulation to fill all the joist spaces along the wall. At the same time, cut an equal number of furring strips. Remember not to try to cut insulation or furring strips on top of the vapor barrier.

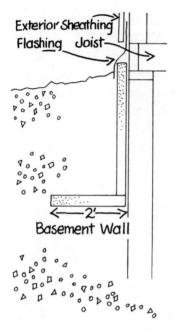

Figure 69 Thorough external insulation of basement and foundations.

To install the insulation, nail the furring strips over the insulation as shown. Be certain that the attached vapor barrier is facing you. Again, repair any rips in it with duct tape. Repeat the procedure for all the exterior walls. Finally, weight the lengths of insulation with some pieces of 2×4 to keep them against the wall.

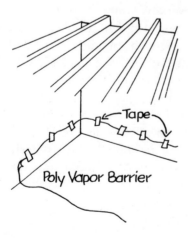

Figure 70 A crawlspace used to distribute hot air must be insulated. Start by laying down a vapor barrier.

Retrofitting Insulation 121

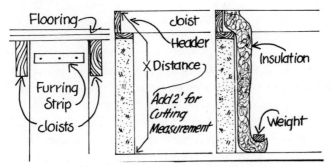

Figure 71 Insulating the masonry walls of a crawlspace.

Internal Insulation of Basements. To overstate the situation a bit, if moisture is rigorously excluded from the basement, then any reasonable scheme for insulating basement walls on the inside will work, but in a damp basement all methods will fail. In dry basements we recommend using foam boards because they have high R per inch and go up quickly. Some foams are available in 4 × 8-foot sheets. One very simple approach (pictured in Figure 72) is to glue the foam boards to the walls with any of a number of suitable adhesives (for example, Sta-tite or Big Stick). Although some of the adhesives are available in hand-squeeze tubes, it is much quicker and neater to use cartridges and a caulking gun.

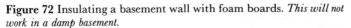

Figure 72 Insulating a basement wall with foam boards. *This will not work in a damp basement.*

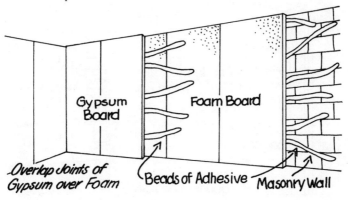

Parallel beads one foot apart should be enough to anchor the boards securely. Taping the joints with duct tape may not be necessary, since a minuscule amount of moisture getting into the masonry wall from the inside is not apt to be harmful. However, taping will not hurt and takes almost no time. Let the adhesive set for several days to make sure it works. In the final step, cover all insulation with ½-inch gypsum board as a fire barrier. The gypsum board can also be installed by gluing it with the same adhesives. If you look around, you may find foam boards with gypsum board already attached.

There is a point of uncertainty here—which is that the glue may fail to hold the gypsum board in place in case of fire. Therefore, secure the board at the top by nailing a batten to the joists so that it clamps the board firmly against the wall, as shown in Figure 73. This works for two walls. For the other two, devise clamps, preferably metal, which can be fastened to the sill and at the same time hold the gypsum board in place.

A modification of the above technique, involving a little more work, can provide greater safety (see Figure 74). Work begins by installing pieces of blanket insulation, with attached vapor barrier, at joist ends. Following installation of blanket insulation, 2×4s are

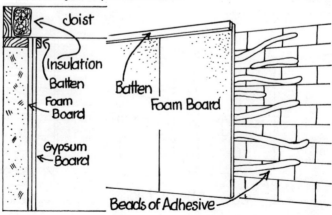

Figure 73 Beadboard and gypsum board in basement wall insulation held in place by batten boards.

Retrofitting Insulation

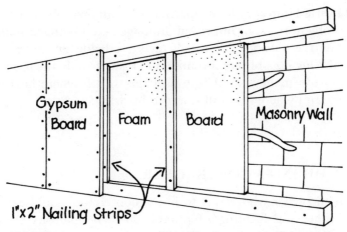

Figure 74 A very secure way of installing basement wall insulation. The foam boards are protected from fire by the well-anchored gypsum board.

nailed to the top and bottom of the masonry walls with masonry nails. One-inch foil-faced urethane or isocyanurate foam boards are then glued to the wall between the 2×4s, foil toward you. Use the same gluing technique described above. Now, 1×2 nailing strips are nailed over the insulation flush with the 2×4s. Masonry nails at least 3 inches long should be used. Addition of these nailing strips will allow a much better installation and finish layer of ½-inch gypsum board which must be used to cover the foam. The trouble with this method is that the 2×4s are subject to rot if water gets through the wall from the outside.

Standard stud walls filled with fiberglass can also be built inside basement walls. More labor is required than in putting up foam boards, but the cost of materials is less for equivalent R-value. Here, again, the studs act as thermal bridges, which is a liability. Should a moisture problem develop after insulating, fiberglass, wood and gypsum board will be degraded faster than foam.

We advise not insulating a basement wall which water penetrates, not only because the insulation will suffer, but also because once the wall is insulated water within it or just outside it may freeze, expand and crack the wall. Damp basements simply cannot be converted into cozy living quarters without first solving the moisture problem through improved drainage and a better seal against water on the outside. The rule is the same below ground as above ground. One possible compromise is to insulate only the top part of a damp basement wall. The top of the wall is where heat loss is greatest and where the wall is apt to be driest.

HIRING A CONTRACTOR

The homeowner who wants to hire a contractor should expect to spend some time locating a reputable one. The majority of insulation contractors (and other home improvement contractors) are honest businessmen, *but* there are companies and individuals who will hire themselves out for good money and do inadequate work, or no work at all.

Where does one begin? With an open mind. *Do not jump to conclusions for or against a contractor without having plenty of information upon which to make a judgment.* Taking the lowest quote on the work you want done will save you money initially, but will it really save on your heating bills? On the other hand, don't assume that the highest bid necessarily means the best job. You could be paying for inefficiency or higher company profits.

Choosing a Contractor. The first information you'll need to collect is a list of insulation contractors working in your area. One source is the local newspaper ads, but remember that anyone can run an ad. The same is true of the yellow pages in the phone book, but they do represent a longer term investment that would not be of much help to the out-and-out fraudulent operator who is in town one week and out the next. Still, a yellow pages listing is not a guarantee of a satisfactory installation. Neighbors who have had insulation installed could provide names, but can they judge whether or not the work was done well?

You can also try local government agencies, such as the FHA offices, state energy office, etc., for listings and perhaps information about companies. Finally, local reputation will be the best guide.

After you have a list of contractors, you should try to get the following information:

1. Ask each contractor for the names of three or four customers for whom he has done an installation. Contact these people and ask them whether they feel they have gotten a good installation, and *why* they feel that way. Do they know how much insulation they actually got? Is their home anything like yours? Did they check out other contractors?
2. Ask the contractors on the list for an estimate of both the cost and the time needed to complete the installation. Have him detail the actual work involved. *Give each contractor the same job specifications!* If you don't, there will be nothing to compare. Have each contractor reply in writing to your questions.
3. Ask the contractor about his liability insurance and the damages and injuries it will cover. Most contractors have liability insurance that covers injuries to the occupants during the job, but only a few have extended coverage that protects you against, say, an allergic reaction to formaldehyde fumes after the job is completed.
4. Check with the Better Business Bureau to see if there have been any complaints lodged against the contractors you have listed. What were the complaints? What does the contractor have to say about them?

At this point, you probably eliminate some names based on cost, bad reputation, or both. If your choice is still something of a toss-up, you might want to consider choosing the contractor who has been in business longest. Business longevity is one kind of community approval of work that is done. Finally, if all other objective reasons for choosing a contractor have been exhausted, you'll just have to go by intuition, choosing the one you can work with most easily.

The Contract. Once you've chosen a contractor, the two of you will need to draw up a contract which clearly delineates the responsibilities of each, including:

1. The type and amount of insulation to be installed.

2. Guarantee of contracted R-value. (This is always difficult to assess. Try to be as precise as possible.)
3. Who will do what work (preparation, cleanup, etc.)?
4. Who is liable for damage to the home and/or injury to the workers?
5. Payment for services.

Most contractors will want you to make a downpayment before work begins and pay the balance at completion or maybe within thirty days of completion. Most will be satisfied with a 25 percent downpayment, and in any case, you shouldn't pay more. You should try to get some guarantee from the contractor that within a reasonable time after receiving the downpayment, he will begin insulating or refund your money. Two weeks is probably reasonable for both of you.

Get a receipt for all monies paid, and a record of services rendered.

These recommendations might seem overwhelming or unreasonable. They aren't. Too often, in all walks of life, serious misunderstandings occur because we believe that the other guy understands exactly what we mean. This is seldom the case, and misunderstandings between friends can be difficult to patch up; how much more so between contractor and client. State explicitly what you want and have the contractor state explicitly what he will do. Live up to your end of the contract, and don't be afraid to ask him to do the same. Enough preaching. . . .

Checking the Work. It is important, or you should feel that it's important, to know that you are getting your money's worth from the contractor. Check the work yourself. In areas where you can make a visual inspection of the retrofit installation (some attics and basements), assessing the quality of insulation isn't too difficult if you've read the sections of this book dealing with those areas. You can tell pretty quickly whether or not insulation is falling out from between the joists in the basement. You can also get a fairly accurate idea of the adequacy of loose-fill in the attic by keeping track of the number of bags used and then calculating the total R-value from the information on the bag and the area of the attic.

NOTE TO CHAPTER V

1. "Material Criteria and Installation Practices for the Retrofit Application of Insulation and Other Weatherization Materials." U.S. Department of Energy Technical Report DOE/CS-0051. November 1978. Page 17.

CHAPTER VI

Installing Insulation in New Buildings

Dana Zak

Those building new homes have the opportunity to insulate them adequately and correctly from the bottom up and avoid retrofitting altogether. It costs less to insulate while building than it does to do it later.

EXTERNAL INSULATION OF THE FOUNDATION

The first step to achieving good insulation values for new structures is to provide good drainage. If water stands around foundations, basement walls, and slabs, chances are it will find its way into the home. This increases the likelihood of cracked foundations and decreases the effectiveness of insulation.

As we suggested in Chapter V, external insulation of heated below-ground concrete structures has a number of advantages:

1. It allows the masonry wall to contribute to the thermal mass of the home. Increasing the thermal mass means increased heat storage and decreased temperature fluctuations.
2. It moves flammable materials out of the home.
3. It increases usable space.

Perimeter and Below-Slab Insulation. In the new cellar hole, after sand or gravel fill is in place, a separate water barrier should be

Installing Insulation in New Buildings 129

installed below the concrete slab basement floor and outside the walls (Figure 75). Polyethylene film, 6 mils or thicker, is the material of choice for the water barrier. In those thicknesses, it resists tearing and puncturing, and is quite impervious to water. If the foundation should crack, the polyethylene sheet will maintain its integrity and still keep water out. In this respect it is superior to any kind of waterproofing tar.

The polyethylene sheets should be as large as possible. If joints are necessary, the overlap should be at least 6 inches and should shed water to the outside just as shingles do. Use duct tape liberally for a perfect seal.

Once the water barrier is in place, and the basement walls are up, insulation can be laid against the walls from the outside. Extruded polystyrene is widely used for below-ground applications, because of its high R-value, imperviousness to rot and water, and resistance to crushing loads. Beadboard, being somewhat permeable to water, is subject to frost damage. Tongue-and-groove foam boards are available and make a neater job than unmatched boards. The boards can be held in place by the pressure of the earth against them.

Regardless of whether you insulate the basement walls fully or not, installation is finished by making certain that flashing covers the top of the foam to prevent water from running behind it and to protect it from sunlight, which can cause foam to deteriorate. Covering can be accomplished in a number of ways; probably the easiest and best is simply to run flashing down to the back-fill. Another, more costly, approach is to apply a brick veneer. For stability the veneer wall must be secured to the basement wall by driving

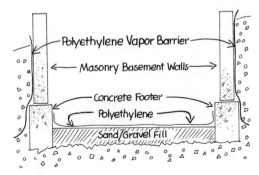

Figure 75 Polyethylene is put in place as a vapor barrier below concrete basement slab before the slab is poured, and outside basement walls.

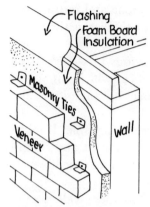

Figure 76 Basement wall insulation outside may be covered with a layer of brick veneer.

masonry nails through the insulation as shown in Figure 76. Incidentally, basement windows, like all windows, are a thermal liability. Why not omit them entirely in your house?

There is not much point in insulating beneath the basement floor, because downward heat loss doesn't amount to much. Recalling the Basic Equation in Chapter II, heat loss is proportional to a temperature difference, and in this case the temperature difference (and therefore heat loss) is always small. Typically, the ground temperature at this depth is around 55°F. summer and winter.

In homes without basements where the concrete slab will serve as part of the floor structure, perimeter insulation needs to be installed as shown in Figure 77. The foam boards should extend inward from the footer by 2 feet, since otherwise there would be a large heat loss in winter through the foundation to the cold ground outside. Note that the insulation is now on the inside of the foundation wall. If it were outside, the foundation would act as a cold bridge and allow a significant loss of heat. Insulate fully under the concrete slab in especially cold climates or wherever the concrete slab absorbs and stores solar energy.

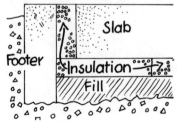

Figure 77 Insulating a concrete slab with foam boards.

INSULATING OVER UNHEATED CRAWLSPACES AND BASEMENTS

Those not intending to heat basements or crawlspaces can help keep upper floors warm by insulating between the floor joists of the lowest heated floor. Before actually building the floor it is wise to plan to use solid bridging instead of cross-bridging, for the following reasons:

1. You'll save time insulating.
2. You'll do a better job of insulating.
3. Your floor will be stronger. The U.S. Forest Service believes that cross-bridging adds little, if any strength to floors.[1]
4. Some people find installing solid bridging faster and easier than installing cross-bridging.

For the money, batt/blanket insulation is the most effective type to use for insulating floors.

If there will be room to work beneath the floor, as is the case with full basements, then reverse flange insulation with an attached vapor barrier should be used since it can be installed after the roof is constructed. This eliminates any water damage caused by bad weather. An added benefit of using reverse flange insulation is that its paper backing reduces any chance that the insulation will sag or fall.

If work cannot be done from below, then standard flange insulation can be installed from above, provided that it can be covered immediately with some sort of weather barrier.

Both types of installation should be done by driving staples every 9 to 12 inches. Tuck up insulation ends against the header as shown in Figure 78. If you've decided to use foil-faced insulation instead of

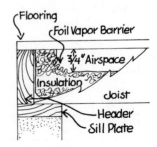

Figure 78 Floor joist cut away to show fiberglass insulation installed with adequate air space between foil vapor barrier and flooring. If the foil touches the floorboards its value as an insulant is lost.

a polyethylene vapor barrier, be sure to allow at least a ¾-inch air space between it and the floor above to get full value from the reflective layer. This is one place where the air space and reflective layer are at least as good as the equivalent thickness of fiberglass.

SILL SEALER

Moving up from the basement, the next area requiring some form of insulation is the sill (Figure 79). Sill sealer installed between the masonry walls and the wood sill plates assures a tight, non-drafty junction. Various sealers are available, but the best choice is mineral wool or fiberglass rolls, which come in just the right width. Installation is quite simple: after unrolling it to size along the top of the masonry wall, cut the wool or fiberglass with a utility knife and push it down over the anchor bolts. It is good practice to cover the sill sealer with a strip of polyethylene to keep rain from soaking it during construction.

INSULATING WALLS

Careful preparation and execution are essential for building and insulating walls. A wide range of insulants and construction techniques are available, and choosing the right combination is not always easy.

One thing to consider is the location of plumbing and wiring. In the past, it has seemed convenient to put plumbing and wiring in outside walls. It is preferable, however, that they be run in inside walls, for the following reasons:

1. This reduces or eliminates the possibility of water freezing in the pipes.

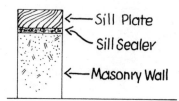

Figure 79 Sill-sealing insulation in place. Installed during house construction, it can save a lot of caulking later on. Try to keep the sill sealer dry during construction. If it does get wet, let it dry before sealing the wall.

Installing Insulation in New Buildings

2. Wiring is removed from an area where it could be completely encased in thermal insulation—a situation which might cause overheating.
3. Putting wiring in inside walls makes for tighter outside walls, since there is no need to cut a hole in a vapor barrier for an outlet, or to partially insulate behind the outlet.

If for some reason wiring must be run in the outside walls, consider using track wiring, which can run along the baseboards, thereby maintaining the integrity of the wall itself.

Insulating Cavities in Stud Walls. Unfaced batt/blanket insulation is the choice for insulating cavities in stud walls. It is possible to use polystyrene beadboard for the spaces between the studs, but it takes too much time to obtain a snug fit unless the studs are spaced with perfect regularity. Beadboard also costs more than fiberglass or mineral wool batts and blankets.

After you have cut the insulation to size, friction will hold it in place between the studs. If you have made the mistake of putting wiring and plumbing in exterior walls, insulation will have to be installed on the outside of them, as in Figure 80. If fitting one piece of insulation behind the wiring and plumbing is impossible, a separate piece will have to be cut and stuffed in.

After stud cavities are insulated, pack around window and door openings. Remember to insulate in the spaces between the joists where they intersect the top plate.

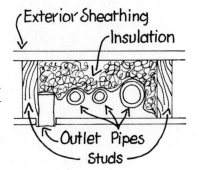

Figure 80 Confining pipes and wiring to inside walls spares you having to stuff insulation around them.

The ideal use of foam boards in conjunction with stud walls is on the inside surface. One inch of polystyrene, urethane, or isocyanurate in conjunction with batt/blanket insulation between the studs can give a high degree of insulation in a small space (up to R-19). There is the added benefit of partially insulating the studs and thereby reducing cold-bridge losses. All foam boards must be covered with a vapor barrier and at least ½ inch of gypsum board to protect them against fire.

Installing a Separate Vapor Barrier. After all insulation is in place, stretch a 6-mil polyethylene vapor barrier across the wall. Begin in an upper corner, and use the top edge of the film against the joist above as a guide for a level installation. Staple every 9 to 12 inches, pulling the film taut as you go. Staple to all framing members, including around windows and doors. Repair any nicks or cuts with duct tape. If you have to use more than one sheet, overlap the ends by 6 inches and tape the joints. Don't trim excess vapor barrier until you have put up the gypsum board.

Using foil-faced batts and a polyethylene vapor barrier can be a mistake. This violates the rule of putting the lowest-perm layer toward the warm side. One builder who did it this way found water trapped between the poly and the foil. It apparently came from the green hemlock studs he had used.

INSULATING POST AND BEAM HOUSES

Post and beam construction, besides being rugged, is esthetically appealing to those who can afford its rustic charm, as well as to some who can't. The appeal, in great measure, is a result of the exposed structural elements and the workmanship put into them. To cover these features with insulation and gypsum board would be butchery of the highest order. Consequently, other insulating techniques have been developed.

One technique is to apply foam boards to the outside of the house and cover them with exterior sheathing. The first step in this process is to put up gypsum board followed by vapor barrier on the outside of the building—with the finish side of the gypsum board facing in (Figure 81). If the distance between the posts is 8 feet, the joints

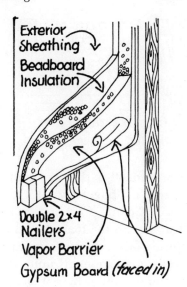

Figure 81 Insulation of a post-and-beam wall can be fast and easy if done from the outside.

between pieces of gypsum board will be behind the posts, minimizing finish work inside. Next, nail horizontal 2×4s to the outside, as shown in the figure, to serve as nailers for the external sheathing and as support for the insulation. Foam boards are then laid up next to the vapor barrier outside, and the sheathing goes down over the beadboard.

FIBERGLASS INSIDE, FOAM SHEATHING OUTSIDE

The recent spurt of growth in the foam insulation industry has led to some innovative uses of foam boards. One is to replace traditional plywood sheathing on the outside of a fiberglass-filled stud wall with 4 × 8-foot foam boards (see Figure 82). Both urethane and isocyanurate are used in this way. The point is to replace a layer of plywood (R-.60) with R-6 or -7. By completely enveloping a house in foam, heat leaks through studs and corner posts are largely blocked. The foam boards may even be more permeable to moisture than the plywood they replace. This is a good thing. Siding can be applied directly over the foam by nailing through into the studs.

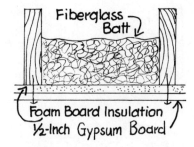

Figure 82 Foam boards used as external sheathing. This is risky unless water leaks are excluded absolutely.

The tradeoff is that the wall must be braced on the inside, since there is no longer plywood to do this job.

The Celotex Corporation has taken another innovative step. In their system the foam boards are faced with aluminum foil to the outside facing a small air space between the foam and siding. This adds a couple more R units, but also puts an excellent vapor barrier on the outside, where it should not be. The problem is solved by leaving a narrow gap between fiberglass on the inside and foam on the outside. This gap communicates with a well-ventilated attic above so that any water vapor that gets into the wall can diffuse into the attic and be carried away through the ventilators. Since the wall is sealed at the bottom, no gross convection can occur within the wall to carry heat away.[2]

It cannot be claimed that this method of sheathing has been tested by the experience of many years, but it does have the nice feature of putting the foam outside and away from fire. Obviously gross water penetration must be prevented, and this depends not only on the quality of the construction but also on continuing maintenance.

INSULATING UNFINISHED ATTICS

Fiberglass and rock wool are the insulants of choice for use in insulating between or directly above ceiling joists. Installation of batts or blankets should be done following the same precautions and clearances already discussed in the appropriate retrofitting sections of this book. Insulation with an attached vapor barrier is preferable because installation of a separate overhead polyethylene barrier is an awkward business, to say the least.

If a second layer of insulation is to be used, observe the following suggestions:

1. Install the second layer without an attached vapor barrier to avoid trapping moisture between layers of insulation;
2. Run the second layer at right angles to the first so that joists are covered and conduction losses reduced.

Foam insulation boards can be used below the batt/blanket insulation if it is desirable not to add a second layer of insulation above the first one of batt/blankets—when electrical wiring is run along the top of the joists, for example, and covering it with the second layer of thermal insulation could cause a fire hazard. Boards are fastened to the joists and *must* be covered with gypsum board at least $\frac{1}{2}$-inch thick to avoid creating a serious fire hazard.

CATHEDRAL CEILINGS

Cathedral ceilings can be trouble in a house requiring high R-values. As in retrofitting insulation in a cathedral ceiling (see page 115), foam boards can be installed above the rafters to almost any thickness. However, this gets quite expensive. On the other hand, the thickness of fiberglass installed between 2×12 rafters is limited to 9 inches (R-30) in order to leave an airway above the fiberglass. Thus, to achieve higher thermal resistance without using wider rafters, one must use a combination of foam and fiberglass. An R-38 roof could be achieved with 3 inches of urethane and 5 inches of fiberglass between rafters. The standard 6-mil poly vapor barrier can be applied from below, but it is much easier to cover the rafters with foil-backed gypsum board.

NOTES TO CHAPTER VI

1. L. O. Anderson, *Wood-Frame House Construction.* U. S. Department of Agriculture, Handbook No. 73, 1970, page 29.
2. For more on ventilating the wall only at the top, write to The Celotex Corporation, 36th and Grays Ferry Avenue, Philadelphia, PA 19146.

CHAPTER VII

Coping with Other Energy Losers

Larry Gay

In many households wrapping the hot water tank with extra fiberglass or a foam blanket is one of the easiest ways to save a significant amount of money and simultaneously enjoy the warm feeling that comes from doing a patriotic act. Heat losses through the jacket of conventional electric water heaters amount to about 17 percent of their total electrical energy consumption.[1]

INSULATING HOT WATER TANKS

For the typical 50-gallon tank a 17-percent heat loss is equivalent to a loss of 83 gallons of oil per year.[2] Jacket losses from conventional gas water heaters are greater because as a rule they are less well insulated. The question is: how much additional insulation is economically justifiable to stem this loss? The answer is: at least 6 inches of fiberglass or the equivalent R-18 foam blanket in virtually all cases, provided the heat lost is totally wasted and does not become useful space heat. In more complicated cases, heat lost from the water heater may become useful space heat in winter, but an extra burden on the air conditioner in summer.

Before discussing the details of insulating hot water tanks, we should note that there are other, often more important, ways of saving energy and money in connection with heating water for domestic purposes. One is by switching to another energy source—such as wood or direct solar energy, both of which are better suited to water heating than are gas and electricity. Gas and electricity are premium forms of energy that should be reserved for the jobs only they can do. Whether the wood and solar alternatives save money today depends very much on region. Where plenty of wood is available, it is the lowest cost option. Solar energy is more problematical, but it can now be justified economically in regions where gas and electricity are relatively expensive.

Even where gas and electricity are the energy source of choice, savings are possible through tactics such as thermostat setback,[3] switching to a smaller heater, preheating water before it goes to the gas or electric tank, and technical improvements of gas water heaters.

Economical Water Heaters. Energy for preheating water can come from wood, from solar collectors, from the air in summer or from the heat in waste water from bathtub, clothes washer or dishwasher reclaimed through use of a heat exchanger on drain lines. This recycled heat is put into incoming water from the city main or well. In a careful study it has been estimated that a preheat tank that simply allows cool incoming water to equilibrate with hot air in the attic in summer can save the typical family 5 percent of the total energy now consumed in water heating.[4] The same study indicated a 13 percent saving in energy by installation of a simple tank-type heat exchanger on the drain line of the kind diagrammed in Figure 83. Assuming 4 cents per kilowatt-hour (kwh) for electricity and an 11-year life span for the system, a life-cycle saving of $233 was calculated for the drain-line heat exchanger. In 1979 the price of electricity is approaching 10 cents per kwh in the Northeast, which corresponds to a life-cycle saving closer to $500. Clearly a heat exchanger on the drain line is a good investment when the price of energy is high. Forward-looking laundry and restaurant owners have already discovered this.

There are many technical changes to gas water heaters that can make them more efficient. Manufacturers have in fact already implemented a number of changes which have resulted in a 17 percent reduction in energy use for the typical gas heater sold in 1976 as compared to 1972.[5] This reduction was achieved by adding insula-

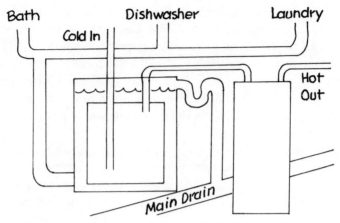

Figure 83 Tank-type heat exchanger in drain line to preheat water going to a gas or electric water heater.

tion, improving baffling in the flue, and changes in the pilot light. By 1980 another reduction in energy consumption of gas water heaters is anticipated by adding more baffling in the flue, forced draft, electronic ignition, and automatic closing of the flue when the gas burner is not on.

Another energy-saving alternative is the use of instantaneous water heaters in place of the standard storage tank heater. These have been standard in Europe for years, but not in the United States where energy was cheap. They heat water only as needed at the point of use and therefore eliminate both heat losses from long pipe lines and standby heat losses from a tank. Where only small amounts of hot water are needed, the instantaneous water heater can represent a saving. It also makes sense in combination with a wood or solar preheater. Both gas and electric instantaneous heaters are now available on the American market.[6] One big drawback to the electric model is that it cannot be switched off by an automatic timer during peak demand hours and still supply hot water, as a storage-type heater can.

Standby Losses from Hot Water Tanks. In Figures 84, 85 and 86, we can see jacket heat loss rates from 20-, 50-, and 80-gallon water heaters at three temperatures as a function of thickness of fiberglass insulation having a density of $\frac{1}{2}$ pound per cubic foot. Although

Figure 84

Figure 85

Figure 86

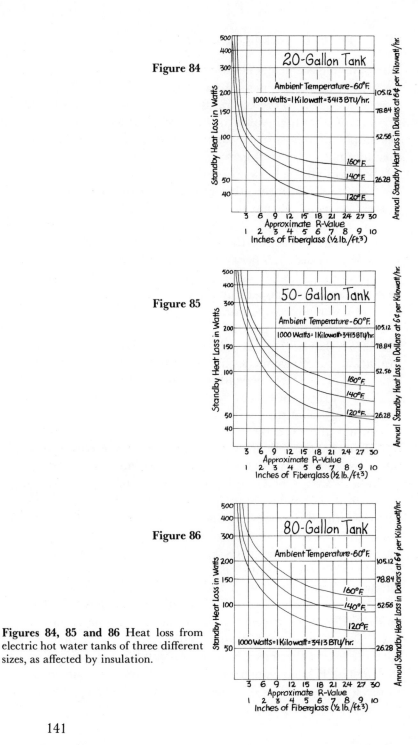

Figures 84, 85 and 86 Heat loss from electric hot water tanks of three different sizes, as affected by insulation.

based on an electric model, these calculations are also applicable to gas heaters for our purposes. In addition to the jacket loss, there is a large loss of heat to the central flue of gas water heaters (about 38 percent of the fuel value of the gas burned in conventional heaters),[7] but this is irrelevant to the question of insulation, and therefore not considered here.

The 20-gallon tank was assumed to be of the lowboy type, standing 2 feet high. Both the other tanks were assumed to be 5 feet tall. It was also assumed that no extra insulation beyond the standard 1-inch factory-installed layer was added on top because of the difficulty in doing so. Furthermore, heat loss through the connecting pipes, pressure-temperature relief valve and drain valve were neglected altogether. These are considered separately below. In all cases an ambient temperature of 60°F. was assumed as a typical average basement temperature.

There are three important points to notice about these curves. The first is that heat loss is reduced dramatically by the first inch of insulation; successive inches bring diminishing returns. The second point is that heat loss depends very much on how big the tank is. And the third point is that heat loss depends heavily on the temperature setting of the thermostat.

To quantify this a bit, the standby heat loss rates from 20- and 80-gallon tanks at 140°F. (and with 1 inch of fiberglass insulation) differ by 143 watts (488 BTU/hr). Translated into dollars, this means a difference of $75 per year if the cost of electricity is 6 cents per kwh, or $21 per year if the cost of gas is 25 cents per therm.[8] Likewise, the difference in standby heat loss for a 50-gallon tank maintained at 160°F. compared to the same tank at 120°F. (with 1 inch of fiberglass) is 104 watts (353 BTU/hr), which amounts to $55 per year for electricity at 6 cents per kilowatt-hour or $15 per year for gas at 25 cents per therm. As a rule of thumb, turning down the hot water thermostat by 20°F. will result in an energy bill reduction on the order of 10 percent for conventionally insulated tanks.

Obviously the general strategy should be to use the smallest tank with the lowest thermostat setting you can live with. But if a small tank at 120°F. is not sufficient, is it better to raise its temperature or to use a bigger tank with a lower thermostat setting? The answer can be found in Figures 84-6.

As Table 6 shows, the heat loss rate from a 50-gallon tank at 140°F. is about the same as for an 80-gallon tank at 120°F. for all thicknesses of insulation. However, for 1 or 2 inches of insulation, heat loss from a 20-gallon tank at 140°F. is less than from a 50-gallon tank at 120°F. The former is therefore more economical in the long run. But, insulated with 6 inches of fiberglass, the 50-gallon tank at 120°F. is just as economical to run. Both tanks will give you enough hot water for two fairly short 110-degree showers back to back. The difference is that the 50-gallon tank will still contain enough hot water when the showers are over for washing the dishes.

One benefit of using as low a temperature as possible on your water tank is that the corrosion rate is slowed so that tank life is prolonged. However, not all household appliances give satisfactory performance by some standards with 120-degree water. The automatic dishwasher will not leave your dishes spotless unless the water temperature is 140°F. It is the only major appliance for which 120°F. is not satisfactory.

The Economics of Insulating. Adding extra insulation on top of factory-installed insulation on water heaters is cost-effective in almost all cases. An exception is the tank that is nearly worn out and needs replacing in the near future. Another exception is where the heat lost from the tank is used to take the chill off the bathroom—that is, to supply space heat. In a careful analysis, one must take into account things such as whether the heat loss places an extra burden on the air conditioner in summer; whether a low off-peak electric rate is available; and whether the local water is hard or soft (this affects life expectancy of the tank).

TABLE 6
Heat Loss Comparisons of Water Heaters (Watts)

Inches Fiberglass	50-gal 140°F	80-gal 120°F	20-gal 140°F	50-gal 120°F
1	175	175	100	140
2	120	120	73	93
4	87	87	58	63
6	75	75	53	55
8	68	70	51	50

Consider first a 50-gallon glass-lined electric water heater maintained at 140°F. Assume it is new and comes with 2 inches of fiberglass (R-6) insulation. Further, assume its life expectancy is 11 years, which is perhaps a bit longer than average.[9] In hard water regions it would be less. Assume all heat lost from the tank is really wasted and that the tank is in a basement where the temperature is 60°F. year round. Finally, assume electricity costs 6 cents per kwh.

Figure 85 showed us that adding a standard blanket of R-11 (3½ inches) fiberglass will reduce heat loss from 120 watts at R-6 to 76 watts at R-17. If the installation cost is $25 the life-cycle saving will be:

$$\text{life-cycle saving} = \underbrace{11\times24\times365}_{\text{hr}} \times \underbrace{(120-76)\times.001}_{\text{kw}} \times \$.06 - \underbrace{\$25}_{\substack{\text{installation}\\\text{cost}}} = \$229$$

A good investment! In fact, it's hard to think of a better one unless it's more fiberglass. It is to be emphasized that this is a conservative calculation because, for simplicity, it ignores the fact that the price of electricity is increasing. The life-cycle saving becomes $343 when the price of electricity is allowed to increase at 5 percent per year. If one prefers to think in terms of payback, the payback time on the investment is about one year.

Repeating the same calculation—using extra insulation of varying thickness—generates the lower curve in Figure 87, which again shows diminishing returns for each additional inch of fiberglass or each additional R-unit. Six additional inches, for a total of eight, can be justified in this case, but the extra trouble and extra space requirement of going beyond makes any further insulation questionable. Even if one can take advantage of an off-peak electric rate of 2 cents per kwh, the life-cycle saving is still $60 with 6 additional inches of fiberglass.

If the tank were 5 years old (6 remaining years of life) and had only a 1-inch layer of fiberglass, there would still be life-cycle savings to be realized by adding insulation; see the upper curve in Figure 87. By inspection one can see that 6 extra inches of fiberglass (R-18) would be justified also in this case, but it would hardly be

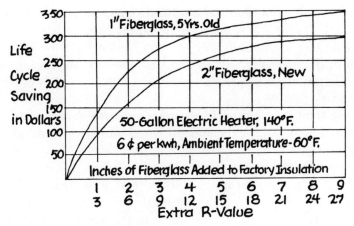

Figure 87 Adding extra insulation to a conventional water heater can save over $300 during the life of the tank.

worth the trouble to go any farther. As a point of comparison it may be noted that manufacturers of solar collectors, who as a group tend to count every BTU, often recommend a minimum of 6 inches of fiberglass on their tanks. The conclusion of a National Science Foundation study in 1974 was that 4 extra inches of fiberglass were justified with electricity at 4 cents per kwh.[10] This agrees very well with our conclusion.

As a final example, consider a 40-gallon gas water heater maintained at 140°F. with 1 inch of fiberglass. Addition of a 6-inch blanket of fiberglass will reduce the heat loss rate from 152 to 64 watts, a reduction of 88 watts or 300 BTU/hr. If the life expectancy is 11 years and the price of natural gas is 30 cents per therm,[11] then the life-cycle saving would be:

$$\text{life-cycle saving} = \underbrace{11 \times 24 \times 365}_{\text{hr}} \times \underbrace{300}_{\substack{\text{BTU} \\ \text{hr}}} \times \frac{\$.30 \text{ in}}{(.50) \times (10^5 \text{ BTU})} - \underbrace{\$25}_{\substack{\text{installation} \\ \text{cost}}} = \$148$$

efficiency factor[12]

Once again a 6-inch blanket of fiberglass means a significant saving, although low compared with the electric tank above because natural gas is still comparatively cheap. *It is hard not to conclude that you should add at least 6 inches of fiberglass to almost any water heater in good condition.* This is especially true of water heaters installed in cold garages. If the fuel were bottled gas (propane), the conclusion would still be valid, since its price is intermediate between that of natural gas and electricity.

If the total thickness of fiberglass is at least 2 inches, there is no advantage to wrapping the tank and insulation with aluminum foil, because virtually all the resistance to heat flow is in the fiberglass. In other words, the resistance to heat flow offered by the jacket-air interface is an insignificant part of the total, so increasing the resistance at the interface with a shiny surface has no practical effect.

Incidentally, before leaving hot water tanks, it's worth noting that boilers are very similar. They are also hot water tanks maintained at 180°F. all year long in the basement. By inference from all we've done so far, the one or two puny inches of insulation on them is not enough.

Installation Details for Electric Heaters. When working with fiberglass or any form of mineral wool, it is wise to wear a nose guard and gloves. Cut fiberglass or rock wool blankets (batts are stiffer and harder to work with) to the correct size using a large pair of shears and then wrap the pieces around the tank (Figure 88) using duct tape to fasten the joints. If there is paper or foil backing on the fiberglass, this should be on the outside to make a tight envelope. Make sure the insulation touches the floor. If you can, add some insulation to the top and make sure there is a good seal between top and side blankets. Cut out ports so that there is easy access to heating elements and thermostat. Be sure that the discharge pipe from the pressure-temperature relief valve is in no way obstructed and that the valve itself is exposed so that operation of its mechanism can easily be tested. (The lever on the relief valve is for this purpose. Lifting it should allow some water to escape. If the lever cannot be moved, the valve is probably corroded and therefore no longer protecting the system.)

When the relief valve is on top of the tank, the easiest way to deal with it is to extend the horizontal pipe it is connected to so that the vertical discharge pipe is on the outside of the extra insulation (see

Figure 88). However, when the relief valve is on the side of the tank—a less common location—the situation becomes more difficult. One way to handle it is to enclose the discharge pipe in insulation while leaving the opening near the floor unobstructed. Be sure that there is plenty of room around the open end of the pipe and the relief valve itself to facilitate checking the valve, and do not reduce the diameter of the discharge pipe or put a valve in the line. This would be in violation of the Boiler Code, the National Plumbing Code and every other code. It would also be dangerous.

The insulation blankets will have a natural tendency to open up circumferentially, so it's a good idea to use several circumferential straps that hold the whole thing together the way hoops on a barrel do. Duct tape that overlaps a good 6 inches is satisfactory. If you now consider the tank an eyesore and cover it with a flammable material, be aware of the fact that most applicable codes require at least 1 inch clearance between combustibles and protrusions, including the cold water pipe which gets hot during standby.

Installation Details for Gas Heaters. There are several very important points about insulating gas heaters. Do not obstruct flow of combustion air to the burner and do not in any way interfere with flow of air into the draft hood on the flue pipe.[13] In fact it is best not to try to add any insulation at all on top of gas heaters. Look carefully at Figure 89, which comes from the Department of Energy. By all means keep flammable materials away from the flue pipe. This includes foam insulation, of course. If that is not enough of a warning, consider the fact that there is a clause in your homeowner's policy that gives the insurance company an out if you knowingly increase the fire hazard in the house.

Insulating Materials Available. You have basically two choices: fiberglass or foam. Vermiculite has been used, but its R-value is low. One foam wrapper available advertises an overall R-value of 4.76. This is hardly the equivalent of 6 inches of fiberglass, although it is quite easy to install.

Fiberglass will be cheaper than foam, but bulkier and harder to deal with. If you are lucky, you will be able to find some of relatively high density. The standard stuff has a nominal density of $\frac{1}{2}$

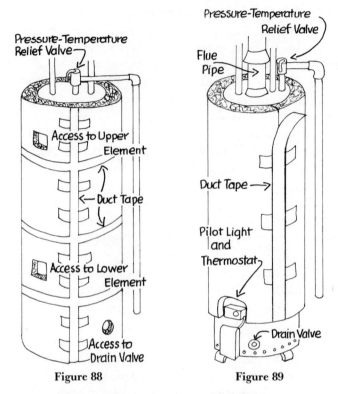

Figures 88 and 89 Electric (88) and gas (89) water heaters shown wrapped with fiberglass blankets. Note pipe from relief valve unobstructed by insulation. In the gas heater, don't let insulation block flow of air to burner at the tank's bottom.

pound per cubic foot. Denser fiberglass has a higher R-value (see Figure 50 in Chapter IV). One-pound fiberglass has an R-value about 16 percent higher than ½-pound fiberglass. The R-value of 2-pound fiberglass takes a similar jump in comparison with that of 1-pound fiberglass. The trouble is, as fiberglass gets denser, it also gets stiffer, and therefore is not as easily shaped to fit the hot water tank. Sears Roebuck sells a retrofit insulation kit using fairly dense fiberglass with an R-value of 4 per inch.

Industry Efforts for the National Cause. Manufacturers of water heaters have already made many of the easy thermal improvements. Some have been mentioned in connection with gas heaters. Most manufacturers now make electric tanks with 2 or 2½ (nominal, sometimes actual) inches of 1-pound fiberglass. The A. O. Smith Company (Kankakee, Illinois), for example, offers tanks with 2½ inches of fiberglass and a total R-value of 10, or 4 per inch, somewhat better than the R-3 we have been using for ½-pound fiberglass. This still falls far short of what will pay for itself, but it's the logical first step for the manufacturer, since packing more fiberglass into the same annular ring around a water heater requires no major production changes. The other significant improvement already taken by some manufacturers is insulating the bottoms of electric tanks which until recently were left bare. There have been some minor improvements too, such as reducing heat leaks around the base of the heating element. The next major step for manufacturers will probably be the use of isocyanurate foam. In a 2½-inch annular ring this would have an R-value on the order of 18. Not bad.

Manufacturers cannot increase the thickness of factory-installed tank insulation indefinitely. Remember, the tank must pass through doorways and down cellar stairs. Even so, most tanks could easily accommodate 4 inches of insulation. With isocyanurate foam the R-value would be an impressive 28.

PIPE INSULATION

It is desirable to insulate both hot and cold water lines close to the hot water tank, especially if they are metal and extend vertically from the top of the tank for any distance. In vertical pipes the natural buoyancy of hot water carries heat farther from the tank than in horizontal pipes. Both hot and "cold" pipes get very hot during standby (feel them) and represent a constant and fairly significant heat loss, which has been shown to be on the order of 14 watts for vertically oriented ½-inch copper tubing.[14] This may not sound like much, but amounts to $7.35 per year or about $75 during the 10-year life of a glass-lined tank. By applying a section of standard 1-inch fiberglass pipe insulation this loss is cut in half. Molded urethane pipe insulation does even better. Some new tanks come with so-called heat traps which prevent buoyant hot water from rising into a vertical pipe above the tank (Figure 90).

In the typical basement installation, the vertical tubing makes a right angle turn a few feet above the tank. You will find that the temperature drops rapidly in the horizontal tubing. The general idea is to insulate any pipe that feels hot during standby.

You could insulate the hot water line all the way to the faucets, but, contrary to a good deal of advice one reads, in most cases this would be a waste of effort, since the hot water in the line usually has enough time to cool between draws regardless of the amount of insulation. Someone working around the kitchen who rinses his hands every 15 minutes may want to insulate the pipe so he doesn't have to wait for hot water, but it's hard to make a convincing case. The better strategy is to put the heater as close to the taps as possible, and, perhaps, to use small-diameter ($3/8$-inch) tubing. The latter re-

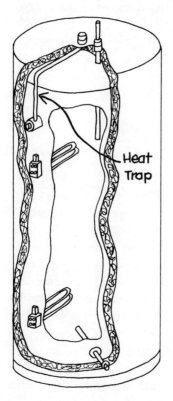

Figure 90 A Sepco electric water heater showing factory insulation (cut away), heating elements, and the "heat trap."

duces the volume of hot water left in the pipe, but also restricts water flow somewhat.

The story is different for pipes that are part of a hot water space heating system and which are more or less constantly hot, depending on whether the pump runs constantly or intermittently during the heating season. If these are in a crawlspace or unheated basement, they should be insulated. There have been enough numerical examples by now that we ask you to accept this on faith.

Insulating cold water pipes in summer is one way to prevent them from sweating. The problem with this is there must be an effective vapor barrier to keep water vapor from penetrating the insulation. According to at least one authority,[15] a simpler way to handle this problem is to wrap some aluminum foil *loosely* around the pipes. This helps to transfer heat to the pipe and to warm it above the dew point temperature of the surrounding air.

In the same vein, pressure (expansion) tanks often sweat in the humid part of the summer. If the tank is in the basement, let it sweat: it acts as a preheater. In fact, a well-designed tank might have fins on it to improve its performance as a heat exchanger. Each gram of water that condenses on the surface of the tank liberates 590 calories. If your pressure tank is effective as a preheater, you may have to cool drinking water in the refrigerator.

INSULATING HOT AIR DUCTS

With typical forced-air furnaces heat loss through uncovered leaders (hot air ducts) in the basement can be as high as 33 percent of the heat generated in the firebox.[16] That is to say, if your fuel bill were $1000 each winter, as much as 33 percent or $330 of it could go toward heating the basement through duct losses, which is all right as long as you want heat in the basement, but a big waste if you don't.

Remember that the R-value of a window is about 1. Thermally speaking, a hot air duct is like a window, except that the still air is on the outside and the breeze is on the inside. True, the metal duct is a good thermal conductor, but that doesn't count because it's the metal-air interface that is the barrier to heat transfer. Thus, a hot air duct also has an R-value of about 1.

If we now add 1 inch of fiberglass (R-3) to the duct, its R-value goes from 1 to 4; heat loss falls by a factor of 4, and so do dollars. That is, we have saved $\frac{3}{4} \times \$333 = \$250$ per year just by adding one inch of fiberglass to the ducts. If the ductwork has an estimated life of 30 years, the life-cycle saving is not exactly peanuts.

Unless you are intentionally heating the basement, insulating hot air leaders will save you a lot of money. Fiberglass duct insulation comes generally in thicknesses of $1\frac{1}{2}$, 2 and 3 inches. The shiny foil surface helps to reduce heat loss by minimizing infrared radiation. The thinner the insulation, the more helpful a foil cover is. There are two warnings: (1) Do not use paper-backed fiberglass or any other flammable material on the ducts; and (2) if your pipes freeze in the basement after you insulate the ducts, you know that you did too good a job. For anyone building a new house, factory-insulated ducts are available.

Table 7 gives the amounts of insulation recommended by the Department of Commerce for ducts in crawlspaces and other places where heat loss constitutes a total waste. An energy cost of 45 cents per therm (100,000 BTU) is assumed as well as a 20-year life for the ducts. For air conditioning, the thicknesses should be about the same for the same number of cooling degree-days wherever a duct is exposed to an excessively high temperature—in a poorly ventilated attic, for example. Remember that a cool duct must have a vapor barrier to keep water from saturating the insulation.

TABLE 7
Recommended Fiberglass Duct Insulation for Unheated Basements and Crawlspaces

2000 heating degree-days	4 inches
4000	5
6000	6
8000	6
10,000	7

Source: S. R. Petersen, *Retrofitting Existing Housing for Energy Conservation: An Economic Analysis,* Building Science Series #64, U. S. Dept. of Commerce, National Bureau of Standards, 1974.

INSULATING REFRIGERATORS AND FREEZERS

Refrigerators are just behind water heaters in residential energy consumption. A typical 16-cubic-foot refrigerator uses about 1800 kwh per year compared to more than 4000 for a 40- or 50-gallon water heater. It is usually located in the kitchen, which means that its waste heat ends up as space heat. In summer, heat from the refrigerator's condenser can make the house uncomfortable or put an extra load on the air conditioner. Ideally this waste heat should be used to heat water. They're working on it.

As with water heaters, there are strategies other than improving insulation to reduce energy consumed in refrigerators. An obvious one is to go back to smaller models without automatic defrost. Table 8 compares various refrigerators and freezers with respect to energy consumption. These figures are for new models encased in urethane foam.

Refrigerator manufacturers have been very quick to adopt foam insulation. For a typical 16-cubic-foot model, filling the 2½-inch space between inner liner and cabinet with urethane instead of

TABLE 8
Daily Energy Use by Refrigerators and Freezers

Refrigerators		
1 door	12 cubic feet and less	1.6 kwh/day
top mount	12-13 cubic feet manual	3.3
top mount	12-13 cubic feet auto-defrost	4.6
top mount	15-17 cubic feet auto-defrost	4.9
top mount	18-22 cubic feet auto-defrost	6.0
Freezers		
chest	14 cubic feet and less	3.3
chest	20 cubic feet and over, manual	4.0
upright	15 cubic feet, manual	4.5
upright	16-17 cubic feet, auto-defrost	6.3

Source: A. D. Little, Inc., *Study of Energy Saving Options for Refrigerators and Water Heaters*, 1977, Vol. I, p. 23. (Available from the National Technical Information Service, #PB-269-153.)

fiberglass gives a 21 percent reduction in total energy consumption and an estimated 10-year life-cycle saving of $128 at 4 cents per kwh.[17]

New refrigerators are very well insulated today. Your concern should be to make sure that you get one surrounded by urethane if you are about to buy a new one, and not spend your time trying to retrofit fiberglass around the old one. For one thing, you could only do it on two sides and the top. The saving just isn't there. Spend your time caulking or building window covers instead.

NOTES TO CHAPTER VII

1. A. D. Little, Inc., *Study of Energy Saving Options for Refrigerators and Water Heaters,* Volume II. 1977, page 28. (Available from the National Technical Information Service, #PB-269-153.)
2. The loss through the jacket is 113 kwh per year. Taking into account the fact that the generating plant is only one-third efficient, this translates into 83 gallons of oil.
3. Setting back the thermostat, like lowering the speed limit to 55 mph, will reduce the incidence of accidents. Between 7 and 17 percent of all scald-burn hospital cases are caused by tap water. Most of the victims are toddlers and preschoolers. Tap water at 138°F. can cause a serious burn if exposure lasts 10 seconds.
4. A. D. Little, Inc., *Energy Saving Options,* Volume II. Pages xxiii, xxiv.
5. See A. D. Little, Note 4.
6. See A. D. Little, Note 4.
7. See A. D. Little, Note 4.
8. One therm = 100,000 BTU; the efficiency of a gas heater is taken as 50 percent here. See A. D. Little, *Energy Saving Options,* Volume II. Page 27.
9. The life expectancy of a cement ("stone") lined tank would be over 20 years.
10. J. J. Mutch, *Residential Water Heating: Fuel Conservation, Economics and Public Policy.* 1974, page 17. (National Science Foundation #R-1498-NSF.) Mutch's analysis includes a discount rate which the present study ignores, since for most people it's around zero. Fortunately the conclusion about economical thickness of insulation on water heaters is not very sensitive to the discount rate, as long as it is within reason.
11. This is not a mistake. The price of gas went up while you were reading the last three pages.

12. See A. D. Little, Note 1.
13. The draft hood's function is to dissipate sudden gusts of wind so they do not blow out the pilot light.
14. W. W. Schultz and V. W. Goldschmidt, "Effect of Distribution Lines on Standby Loss of Service Water Heater," *ASHRAE Transactions,* Part 2. 1978, page 256.
15. G. B. Wilkes, *Heat Insulation.* New York, John Wiley, 1950, page 148.
16. This was the figure found at Twin Rivers Project in New Jersey (*Energy and Buildings,* 1. 1977/78, page 218). Textbook estimates tend to be lower.
17. A. D. Little, *Energy Saving Options,* Volume I. Page xvi.

CHAPTER VIII

Laws, Government Programs, and Codes Affecting Insulation

Eugenia C. Worman

Before the 1970s, few people insulated their houses or work places; fuel was cheap—so cheap that it was more costly to insulate a building than it was to burn fuel. Furthermore, contractors, intent upon keeping down construction costs, were not overly concerned with reducing any structure's daily operating expenses; so buildings went up unarmed against the heat and cold.

Since the oil embargo of 1973, oil has quadrupled in price. Coal has become two-and-a-half times more costly, and this in turn has caused the upward mobility of electricity bills. Natural gas, although still reasonable, has become almost impossible to obtain for new buildings. As price increases and shortages became an everyday concern, the federal government began to act through its various departments and agencies.

First, it spread the word, particularly to civic leaders and building owners, about tightening up existing buildings with weatherstripping and insulation, so that it would take less energy to maintain a comfortable temperature. Then the Department of Housing and Urban Development (HUD) became stricter about the insulation requirements for mobile homes and about storm windows and insulation in other buildings that came under its jurisdiction. Finally, federal buildings came under scrutiny, with the President directing agencies to reduce energy use in all these buildings by 5 percent. The General Services Administration (GSA) published guidelines for saving fuel in federal buildings, and set out design guidelines for any government office about to be built.

It soon became apparent, with costs continuing to increase and shortages threatening, that all-encompassing legislation was needed. It was also obvious that the federal government needed to coordinate its efforts with the states, many of which were already proposing energy measures of their own. States, it was assumed, could draw up conservation plans that best suited their own geography, economy, and climate. What they couldn't do was to carry out adequate programs without additional money and technical assistance.

Therefore, the first federal act to be signed into law in December of 1975 (The Energy Policy and Conservation Act, Public Law 94-163, referred to as EPCA) authorized $150 million to be spent over a three-year period for qualifying state conservation programs. The goal of EPCA was a 5 percent reduction in energy use below the level predicted for 1980.

The State Energy Conservation Program that grew out of a section of this law and another similar one passed a year later (The Energy Conservation and Production Act, Public Law 94-385) were sets of rules governing the lighting and insulation of new buildings, new sections of old ones, and renovated structures. In drawing up their codes, states then depended on a standard for insulating new buildings published by the American Society of Heating, Refrigerating and Air Conditioning Engineers (ASHRAE), which contained minimum requirements for "thermal efficiency" or heat retention. States that didn't use ASHRAE 90-75, as the code was called, depended on a subsequent but similar code drawn up by a conference of state administrators and building industry experts—the Model Energy Conservation Code. To get funding, states had to comply with a set of federal guidelines and promulgate recognized thermal efficiency standards.

The State Energy Conservation Program is now run by the Department of Energy (DOE), and all states and territories, except the Trust Territory of the Pacific, have submitted plans, gotten approval, and received assistance. Each state or territory studied its own energy consumption and supply, figured out whether there was any chance of reducing the demand on fuel by 5 percent of what was projected for 1980, and then, assuming there was, wrote a document spelling out how the goal would be reached. In this, they worked with federal administrators, who in turn decided how much money would be needed.

Codes and standards for new building construction are specifically the business of state and local governments. Other aspects of fuel conservation are the province of both state and federal agencies: tax credits for the cost of weatherizing, loan programs (largely for the poor and elderly), legal directives that spell out the ways utilities are expected to help their customers assess and finance conservation projects, standards for home insulation materials, and plans to reduce energy waste in government buildings.

These latter considerations are far more important for overall conservation than are the codes for building new structures— because in any given year only two to three percent of all buildings are new. Even if all new houses and public halls are well insulated and equipped with efficient heating and lighting systems, we are still left with a vast number of leaky, fuel-consuming structures in need of retrofitting for energy conservation.

But states vary widely in just what they offer the homeowner who wants to tighten things up. Further on in this chapter we will list the programs several states have enacted; but for the present, we need to look at the newest federal law, since its provisions apply everywhere and affect all of us.

From 1973 to 1977, the federal government worked mainly on conservation guidelines for its own buildings, on state plans, on insulation for low-income housing, and on financing for renewable resource measures (solar, geothermal, and wind) in houses, public buildings, and factories. In November of 1978, The National Energy Conservation Policy Act was signed into law, broadening the government's role as a helper and lender. Here are its programs.

1. WEATHERIZATION

This DOE plan, for low-income families only, provides funds for insulation, storm windows, and other heat-saving materials. The money is administered by local community action agencies (CAA), which buy the materials, oversee the work, and use crews paid for by funds from the Comprehensive Employment and Training Act (CETA). To qualify for help from the DOE in weatherizing, a family's annual income must be at or below 125 percent of the poverty level; for example, the limit for a nonfarm family of four is now $7,750. The CAA determines how much a house needs in weatherstripping, caulking, window repair, roof patching, furnace adjustments, storm windows and doors, and insulation of ducts, floor, walls, attics and crawlspaces. Even the most basic work on a house—just the caulking, storm windows, weatherstripping and repairs—results in a saving of 8 barrels of oil a winter, or about $440 in the North. There is no outlay by the homeowner.

2. RESIDENTIAL CONSERVATION SERVICE

This program directs gas and electric utilities to help their customers save energy: specifically, utilities must do the following:

a.) Hand out lists of suppliers, contractors, and banks or loan associations that have the Energy Secretary's stamp of approval, so that customers can avoid being cheated or being charged prohibitive interest rates.

b.) Offer to inspect customers' houses to determine the cost of buying and installing insulation and other energy-saving measures. Also the utilities should give customers an idea of what they can expect to save as a result.

c.) Offer to arrange to have these measures taken. Utilities can't do the work themselves, however, except for putting in clock thermostats, furnace modifications, energy storage devices, and various mechanisms to control the amount of electricity a household uses.

d.) Offer to arrange financing. Utilities themselves may finance loans that do not exceed $300, and they may also finance any work they do themselves in the categories mentioned in point (c).

e.) Keep records of all money expended on this program separate from the records of utilities' other expenditures. Customers bear the costs of labor and materials for their own projects, but the expenses incurred by the utility in connection with the program are passed along to all customers on the monthly bill.

f.) Permit customers to repay loans (principal and "fair and reasonable" interest) as part of their monthly bill over not less than three years (unless the customer chooses to pay over a shorter time). Also, an outside lender may have repayment made through the utility.

Who is eligible for the Residential Conservation Service? All gas and electric utility customers who live in buildings with four or less apartments.

This plan—the Residential Conservation Service—is not expected to take effect much before June 1980, because, first, the Department of Energy must publish standards to serve as guidelines for the states and territories; then the standards and rules have to be discussed at regional hearings, and perhaps changed in minor ways; next, the states have to incorporate the standards and rules into their own energy conservation plans; and finally, all the state plans must be approved by DOE.

One final word, though. The Residential Conservation Service is not revolutionary in the energy scene; the Federal Energy Administration initiated a Home Energy Savers Program in the mid-seventies, in which states were encouraged to disseminate fact sheets and films about insulation and other home conservation devices. And a few individual utilities have had their own conservation programs running for several years.

Here's what Michigan Consolidated Gas Company did in 1972, for example. In cooperation with other area gas and electric companies, it promoted increasing ceiling insulation to a recommended standard of R-19, and provided low-cost financing to customers who complied. Thirty-two months later, an estimated 200,000 houses had been insulated!

3. RURAL HOUSING WEATHERIZATION PROGRAM

Finally, the Department of Agriculture runs a rural housing weatherization program separate from the Residential Conservation Service. Through its Farmer's Home Administration (FmHA)

Laws, Government Programs, and Codes Affecting Insulation

it provides loans and grants to low-income rural families with an income under $15,600 for weatherizing individual houses. Each loan has a $1,500 limit. The loans are arranged through rural utility cooperatives. The weatherization work, contracted for by the utility, has to meet FmHA or utility standards, whichever are stricter, and again the repayment is made on the monthly bill. Customers have five years to pay up.

If a person's income is too high to qualify for a FmHA loan or grant, another program (section 504) provides loans—limit $5,000—to homeowners who will occupy their houses after the insulation work is finished. Another part of section 504 provides grants for people over age sixty-two who wish to insulate their houses.

OTHER LOANS AND GRANTS

Besides the FmHA encouragements just mentioned there are many other plans carried out by HUD through its local agencies that aim at helping middle- and low-income people to conserve energy. Here are descriptions of these HUD aid programs.

Title I Home Improvement Loans. Operating through local lending institutions, HUD offers insured loans for major or minor home improvements including solar systems and all conservation devices (except free-standing items, such as windmills). Borrowers must have a large enough income to handle loan payments, and must have a record of reliability. The loans are made for 12 percent interest or less, and the maximum amount is $15,000 for single houses, with fifteen years to repay. The limit for apartment buildings is $25,000—twelve years to repay—and $5,000 is the limit for people renting single apartments, though they must have leases that run six months beyond the final loan payment.

Loans of less than $7,500 are usually unsecured personal ones. Anyone interested in taking part in this plan fills out an application, signs a note, and files a completion certificate if the work is done by a contractor.

Section 203 (b) and (k) Home Mortgage Insurance. In 203 (b) HUD helps homeowners get mortgages for single-family dwellings by insuring approved private lenders against loss in case of default by borrowers. This plan encourages home ownership. If a prospective homeowner plans to put in a solar system, the loan for it is insured along with the basic mortgage.

The 203 (k) program provides insurance for loans to improve houses. As energy conservation standards become enforced under the NECPA (National Energy Conservation Policy Act of 1978) this insurance plan will become more useful.

Section 312 Property Rehabilitation Loan. For both low- and moderate-income applicants, HUD provides direct loans at 3 percent interest for improving old buildings—single- and multi-family dwellings as well as nonresidential buildings. Borrowers must show that they cannot get a loan elsewhere on comparable terms. In making loans under Section 312, HUD works through its own area offices, and through local governments and their agencies. The objective is to bring property up to local building standards, which now include basic weatherizing requirements. To make enforcement of building codes feasible, HUD requires the property in question to be part of its own federally assisted plans, such as Urban Homesteading or the Community Development Block Grant Program.

Urban Homesteading. Working with local housing agencies, HUD transfers properties it currently owns to them, on the sole condition that they have an approved homesteading plan. These plans must show that all resources needed for rehabilitation of the former HUD building are available in the given town or city—all the technical assistance, the financing, the municipal services. Then the local government sells the house to a "homesteader" for a token sum (as low as $1). The homesteader must then repair the building to meet local health and safety standards, must live in the building steadily for at least three years, and, within eighteen months of occupying the house, must meet local building code standards. Since weatherstripping, caulking, and putting on storm windows are increasingly a part of municipal building codes, homesteaders are now doing the necessary weatherizing as part of the overall renewal effort. A homesteader acquires full title to the property when all requirements are met. Since homesteaders are dealing with government-assisted property, they are eligible to apply for Property Rehabilitation Loans.

Community Development Block Grant Program. HUD awards "block" grants to local governments for a variety of development activities they themselves have planned and have the manpower to administer. This is a change from the day when there were a number of separate grant programs, such as urban renewal, neighborhood facilities grants, model cities, historic preservation grants, and so forth. CDBG (Community Development Block Grant Program) replaces these individual programs, and has flexible guidelines which allow local governments to determine their own priorities. Many undertake to upgrade housing and expand work opportunities for low-income groups.

Cities and urban counties are entitled to CDBG funds if they demonstrate a certain degree of need, calculated by a formula that takes into account population, poverty, housing conditions and growth lag. Small towns get whatever is left over; however, they are guaranteed a minimum amount based on what they got under former grant programs. Over five years the money gradually decreases.

Section 312 rehabilitation loans, as we have said, can be coordinated with CDBG, and so can other loan possibilities, such as weatherization for the poor, and even Title I home improvement funds.

ENERGY TAX CREDITS

Help from the Departments of Agriculture and of Housing and Urban Development, important as it is, will not reach as many people as federal and state energy tax credits. Congress now provides tax incentives in the form of two types of credit: one for people who install insulation, and another for those who put in renewable energy equipment (solar, wind or geothermal).

Let's take up insulation first.

Credits for Insulation. The Internal Revenue Service leaflet which describes this credit (publication 903: "Energy Credits for Individuals") defines insulation as "any item that is specifically and primarily designed to reduce the heat loss or gain of a dwelling or water heater. It includes, but is not limited to, materials made of fiberglass, rock wool, cellulose, Styrofoam, urea-based foam, urethane, vermiculite, perlite, polystyrene, reflective insulation and extruded polystyrene foam." These are installed in ceilings, walls, and floors, and around pipes, under roofs, or between roof layers, and around forced air ducts and hot water heaters. Decorative items such as drapes and paneling don't qualify for credit.

Other devices for which you may claim credit: storm doors and windows, the weatherstripping around them, caulking, energy-saving thermostats, replacement burners which reduce the amount of fuel consumed in gas- and oil-fired heaters, flue dampers which shut off flow of air from furnace to chimney when the furnace isn't going, meters that show energy usage, and furnace ignition systems that replace pilot lights. Unfortunately you cannot get credit for a heat pump, fluorescent lights, a wood- or peat-burning stove (although, as this is being written, there is a bill before Congress sponsored by James Jeffords of Vermont that would give credit for wood stoves), or hydrogen-fueled equipment.

The tax credit for insulation and the other thermal improvements listed above amounts to 15 percent of what you spend, up to $2,000. You compute this on form 5695 and transfer the amount of the credit to line 45 of your Form 1040 tax return. The credit cannot exceed the amount of tax you owe in any one year, but any excess may be carried over to the next year.

Example: Suppose in a given year you spend $500 on insulation and $700 on storm windows and weatherstripping. Then your credit is 15 percent of $1,200 or $180.

The energy-saving devices you put in your house must be new and have a life expectancy of at least three years. And the house itself must be your main residence, must be located in the United States, and must have been substantially completed at the time the conservation devices were added.

Credits for Renewable Energy Equipment. Now for the expenses incurred in using solar, wind or geothermal energy. Here you claim 30 percent of the first $2,000 spent, and 20 percent of the next $8,000.

Laws, Government Programs, and Codes Affecting Insulation 165

You are eligible from the time the installation is completed. Again, you may spread your credit over the next several years, and again, the house must be your principal dwelling place and located in the United States.

Renewable energy equipment must be new and have a life expectancy of five years. Standards for these properties are about to be published by the Secretary of the Treasury; when they are, they must be met.

It is important to realize that tax credits supplement other government programs. Thus, a grant or loan under another federal plan, such as a Title I improvement loan, does not reduce one's eligibility for credit for either weatherizing materials or renewable energy equipment. Let's say you are going to borrow some money under Section 312 of the property rehabilitation loan program in order to insulate the walls of your house—you still use the money you pay out for materials as the cost basis for your tax credit, and the fact that you are financing the insulating through a government loan does not reduce or in any way affect your eligibility for tax relief.

Many of the other government conservation plans overlap one another. A person signing up for a Title I home improvement loan may find he is also eligible for 203 (b) or (c) mortgage insurance, or that he can use the Residential Conservation Service. Or, a community action group may find itself eligible for two or more programs at once. For example, a Community Development Block Assistance Grant may coordinate with weatherization for the poor, and, at the same time, a Section 312 rehabilitation loan.

RULES FOR LABELING INSULATION

The last government action worth noting is the Federal Trade Commission's proposed rules for the labeling and advertising of insulation. This is a "truth in packaging" move which aims to establish a standard method for determining R-value. This should mean that a customer can discover from a label what insulation meets his particular need, and that he will not be misled by false claims.

Some buildings need material that has a high R-value and some do not, depending upon a number of circumstances such as climate, building design, wall thickness, and so forth. For this reason, manufacturers will now be required to label their packages and provide fact sheets which give: (a.) an explanation of R-values; (b.) the R-value and thickness of the material at hand; and (c.) an estimate of how large an area you must cover to obtain the listed R-value.

The Federal Trade Commission's rules should mean that a nonprofessional can get an idea what insulation best suits his purposes, and how much he will need to spend to get certain results. The rules do not demand that manufacturers explain the relative merits of different types of insulation or that they spell out for the customer how much insulation he needs to install to realize a particular savings on fuel bills. Consumers are encouraged to report violations of rules or any deceptive practices that they run into to the Federal Trade Commission, Pennsylvania Avenue at 6th St. N.W., Washington, D.C. 20580.

STATE PLANS FOR CONSERVATION

Having a grasp of what the Federal Government has come up with in the way of aid for conservation-minded homeowners and renters, we can now look at some of the state plans in detail and their requirements for energy conservation in all new buildings.

We mentioned earlier the use states have made of the building code published in 1974 by the American Society of Heating, Refrigerating and Air Conditioning Engineers (ASHRAE 90-75). More recently another code has emerged, this one resulting from a contract between DOE and the National Conference of States on Building Codes and Standards (NCSBCS, called "Nixbix"). This code differs from ASHRAE 90-75 only in minor ways, the chief one being that it applies to new additions to old buildings as well as to new structures themselves.

Known as the "Model Energy Conservation Code," this code regulates the design of outside walls as well as the selection of mechanical, electrical, and lighting systems to achieve the highest possible degree of energy efficiency. It sets forth minimum requirements for all new buildings, both public and private—the only exceptions being those not heated or cooled by fuel and those with a very slight dependence on fuel (less than one watt per square foot of floor area).

The code specifies the following:

1. Issuance of a building permit depends on the approval of a building official responsible to a government agency who goes over the building plan to see that it meets specific conservation requirements. (These requirements affect the building's design, the materials in the thermal shell, all heating and cooling systems, the rate at which heat passes through the building's outside wall, and the R-value of insulation to be used.)
2. All construction work is subject to inspection by the building official, including a final going-over after all work is completed.

To make the code as flexible as possible, its authors provide three approaches:

1. Design criteria for total energy use of a building, including subsystems which may be using nondepletable energy sources (solar, wind, or geothermal).
2. A plan in which the different parts of the building and its mechanical systems have to meet performance standards. These parts include the building's envelope, the mechanical systems, the hot water supply, and electric power, and light.
3. Requirements for smaller residential buildings—those less than 5,000 square feet in gross floor area, and three stories or less in height.

Twenty standards, listed by number, title, and source, are the pith of the Model Energy Conservation Code, leaving little room for guesswork by the building contractor. An appendix gives detailed diagrams for insulating different sorts of walls—wood stud, steel stud, masonry, brick masonry, and concrete—as well as for roofs, ceilings, and floors. These charts show the allowable R-value and the rate of heat flow through the wall or roof.

How much difference do these conservation building codes make?

Studies of ASHRAE 90-75 have indicated that its adoption and use results in energy savings of 30 to 40 percent in commercial buildings and 11 to 30 percent in homes. The cost of complying with ASHRAE 90-75 has not been significantly different from the expense of conforming with ordinary building codes.

What do building codes have to do with state energy conservation plans? We mentioned before that in 1975, when the Energy Policy and Conservation Act was signed into law, the first requirement for getting federal money was that states have energy-efficiency codes for new building construction. State legislatures reacted in a variety of ways, but mainly they either appointed a body to draw up a code, or authorized in a bill the adoption of a certain code, usually modeled on ASHRAE 90-75.

Except for building codes and standards, many of the state conservation measures overlap those that are growing out of the National Energy Conservation Policy Act. Thus, tax credits have been allowed by some states on their income tax forms; other state legislatures have already required their utilities to give out lists of reliable insulation contractors as well as loan companies with reasonable rates for home insulation loans.

Policies and legislation affecting insulating in the states are in flux at this writing. Any attempt to run down what the fifty states have done would be seriously outdated by the time this book is published. Therefore we'll limit ourselves to a brief look at some of the insulating policies of a few pace-setter states.

Rhode Island. Rhode Island had a well-developed home-audit program by the time the Residential Conservation Service came into being. Two years before the federal government's National Energy Conservation Policy Act, Rhode Island's governor had appointed a citizens' committee made up of volunteers with expertise in the fuel conservation field. Their job was to get a home-weatherizing service underway, and to do this they formed a nonprofit organization. Because the birth of this organization predated any federally-funded state conservation plan, the group's funds came only from the pledges they were able to get from large corporations, banks, and one of the state's large utilities.

Once established, the committee arranged for home audits, to be followed by whatever action the homeowner wished to take. It also facilitated contacts with reliable contractors and bank loans at reduced interest rates. More recently the organization has conducted furnace inspections because they discovered that fuel costs in newly weatherized houses were still high, and the culprit was a poorly maintained heating system.

Oregon. This state, too, has an extensive weatherization service conducted by investor-owned gas and electric utilities. These are required to give out information, inspect houses and make cost estimates, install insulation and other energy-saving items up to a cost of $2,000 for each dwelling, and set up financing at $6\frac{1}{2}$ percent interest. (Oregon gives lending institutions a tax credit to make up the difference between $6\frac{1}{2}$ percent and the current market rate). Although this service bypasses mobile home dwellers, it is available to renters if they have their landlord's consent. Utilities charge their customers according to a formula devised by the state's Public Utilities Commission.

Oregon's thirty public utilities and three-hundred-odd fuel oil dealers are also required to provide weatherizing services to their customers. Although they don't do the work themselves, they still have to tell customers how to get the work done. The utilities also inspect houses, do cost estimates, and determine the best contractor and lending service for a customer's location and circumstances.

Any low-income Oregon resident over sixty who is not receiving federal weatherization help is eligible for assistance from the state of up to $300 per house.

Oregon's veterans can get home loans for houses built before July 1, 1974 only if the house is weatherized up to state standards. They are given 120 days after the issuance of the loan to insulate, to weatherstrip, and to otherwise bring the house up to snuff. The cost of weatherizing is included in the loan's principal.

When it comes to insulation in new houses, some states are much stricter than others. For example, North Carolina prohibits electric hook-up or occupancy of any house built after January 1978, until the insulation installed is inspected and seen to comply with state minimum standards. The inspectors are state trained; the municipal and county governments in North Carolina are charged with certifying plans.

North Carolina gives a tax credit, up to $100, for insulating. Some of the other states that do this (at this writing) are: Alaska, Arizona, Arkansas, Kansas, Montana, Oregon, Vermont, and Wisconsin.

Readers who are in doubt about assistance that may be available through their state governments for insulating work in their homes should inquire of their state energy offices, or—in some states—of their state housing authorities (look under Government, State, in the Yellow Pages).

Notes on Further Reading

The field of home insulating has lain fallow for many years, but now—with the energy crunch really upon us—it's once again being plowed, and furiously. New information is being developed quickly. To a large extent, though, the new information has not yet been incorporated into insulating manuals. For anyone interested in keeping abreast of the latest developments, the *Transactions* and *Journal* of the American Society of Heating, Refrigerating and Air Conditioning Engineers (ASHRAE) are indispensable. These are available in university and technical libraries. Alternative energy publications such as *Solar Age* magazine and *Home Energy Digest* also carry up-to-date articles on insulating. *Solar Age* is a monthly published at Church Hill, Harrisville, New Hampshire 03450. *Home Energy Digest* is published four times a year (address: 8009 34th Ave. South, Minneapolis, Minnesota 55420).

Among recent books, two stand out as especially authoritative: *Insulation Manual*, published in 1979 by the National Association of Home Builders Research Foundation (627 Southlawn Lane, P.O. Box 1627, Rockville, Maryland 20850); and *From the Walls In*, by Charles Wing (Boston, Atlantic-Little, Brown, 1979).

Index

Air ducts:
  insulation, 151–2; **table 152**
Air leaks in house, *see* Heat loss; Infiltration
American Society of Heating, Refrigerating and Air Conditioning Engineers (ASHRAE), 157, 166
Asbestos, 87, 88
Attic:
  insulation in new house, 136–7
  retrofit insulation, 102–15
    with blanket insulation, 105–8; *ill. 106, 107;* second layer, 107–8; *ill. 107*
    blowing in loose-fill, 110–12; *ill. 110*
    hand-poured loose-fill, 108–10; *ill. 109*
    inspection, 103–5
    with knee-walls, 112–15; *ill. 113, 114*
  ventilation:
    amount required, 50

Balance temperature, 16, 17–19
Basement:
  insulation in new buildings, 128–30; *ill. 129, 130*
  quantity of insulation required, 35–6
  retrofit insulation, 119, 121–4; *ill. 121, 122, 123*
Better Business Bureau, 125
Blower for insulation, 110; *ill. 110*
BTU (British Thermal Unit), 8
Building materials:
  permeability to moisture, **table 48**

Cathedral ceiling:
  insulation in new house, 137
  retrofit insulation, 115; *ill. 116*
  use of vapor barrier, 50
Caulking, 20, 21, 57–67
  compounds, 59–60
  foam in aerosol can, 60
  quantity required, 60
  doors, 61–2
  foundation sills, 62–3; *ill. 62*
  gun:
    for affixing foam boards to basement walls, 121
    selection, 59
    use, 61
  tax credits for, 164
  windows, 58–61
Ceiling:
  cathedral, *see* Cathedral ceiling
  insulation:
    clearances, *ill. 106*
    quantity required, 35
Cellar, *see* Basement
Cellulose, 88–9
  blown into existing walls, 116–17
  flammability, 89
  manufacturing process, 9, 89
Cement:
  for masonry pointing, 68–70
Checklist of weatherization, 58, 83
Chimney:
  caulking, 64
  heat loss, 20, 21–2
Cold bridge, 31–2; *ill. 31*
Comfort:
  expectations, 3–4
Community Development Block Grant Program (CDBG), 163

Index

Concrete:
  for masonry pointing, 68–70
  slab:
    perimeter insulation, 130; *ill. 130*
Condensation, *see* Water vapor
Conduction, 6
Conservation plans of states, 166–70
Consumer Product Safety Commission (CPSC), 87
Contractor, 124–6
  checking work, 126
  contract with, 125–6
  selection, 124–5
Convection:
  and dead air spaces, 25
  principle, 6
Cooling balance temperature, 18–19
Cost effectiveness of insulating:
  air ducts, 151–2
  hot water tank, 143–6
  refrigerators and freezers, 153–4
  walls and ceiling, 33–5
  water pipes, 149
Cost of:
  energy, **table 40**
  heat loss, 30–2
  insulating materials, **table 91**
Crawlspace:
  insulation, 119–20; *ill. 120, 121*
Cross-bridging, 100; *ill. 101*
  and retrofit floor insulation, 118
  and retrofit insulation with blankets, 106–7; *ill. 107*
Curtains (thermal), 41, 43, 44–6; *ill. 41*

Damper:
  fireplace and furnace, 21
  tax credits for, 164
Dead air space, 7–8
  and convection, 25
  in pockets within insulating materials, 23–4
Degree-days (DD), 16–19; *ill. 17, 18*
Dept. of Energy (DOE):
  recommendations:
    on loose-fill load-limits in attic, 109
    on retrofit insulation around electric wiring, 103
  research on insulation, 8, 10, 84–5
  State Energy Conservation Program, 157–8
Dept. of Housing and Urban Development (HUD), 156
  programs, 161–3
Dew point temperature, 46
Doors:
  caulking, 61–2
  heat loss, 21
  refitting, 76–7; *ill. 77*
  sills:
    leveling, 74
  storm:
    caulking, 65–7
    tax credits for, 164
  weatherstripping, 70–7
    kit, 73; *ill. 75*
    materials, 70–4; *ill. 71, 73*
    methods, 74–7
    shoe, 74, 76; *ill. 75*
    sweep, 76; *ill. 75*
Door shoe, 74, 76; *ill. 75*
Drainage:
  affects basement insulation, 123–4, 128
  affects R-values, 36
Ducts:
  insulation, 151–2
    amount recommended, 152; **table 152**

Electrical wiring:
  caution in retrofit insulation, 102–3, 105–6
  faulty, 104
  location, 132–3; *ill. 133*
Energy Conservation and Production Act (1976), 157
Energy conservation plans of states, 166–70
Energy Policy and Conservation Act (1975), 157, 168
Energy tax credits, 163–5

Federal Trade Commission:
  regulation of insulation labeling, 165–6

Fiberglass, 85–8; *ill. 86*
  flammability, 87–8
  health considerations, 87
  as hole-filler before caulking, 60
  manufacturing process, 9, 86
  R-value, 23, 24, 86–7; *ill. 86*
  for water heater, 147–8
Fire safety, 51–2
  see also Flammability
Flammability of:
  cellulose, 89
  fiberglass, 87–8
  insulating materials, 51–2
  perlite, 51, 97–8
  polyisocyanurate foam, 92
  polystyrene foam, 51
    extruded form, 90
  polyurethane foam, 51, 92
  urea formaldehyde foam, 51, 95
  vermiculite, 51, 97–8
Floors:
  framing, *ill. 100*
  insulation:
    in new buildings, 131–2; *ill. 131*
    retrofit, 117–18; *ill. 118*
Foam:
  caulking compound, 60
  for insulating water heater, 147
  polyisocyanurate:
    flammability, 92
    manufacturing process, 9, 92
  polystyrene, *see* Polystyrene foam
  as sheathing of new house, 135–6; *ill. 136*
  urea formaldehyde:
    danger of, 93, 98n
    flammability, 51, 95
    manufacturing process, 9, 93
    shrinkage problem, 93, 94–5; *ill. 94, 95*
Foil:
  between radiator and wall, 96–7
  to increase R-value of air space, 25
Foundation:
  insulation in new building, 128–30; *ill. 129, 130*
  sills:
    caulking, 62–3; *ill. 62*
    of garage, 63

  sealer, 132; *ill. 132*
Framing (roof, floor, wall), *ill. 100*
Freezer:
  insulation, 153–4; **table 153**
Fuel:
  conversion table, 39

Garage:
  caulking foundation sill, 63
General Services Administration (GSA), 157
Glazier's points, 79; *ill. 79*
Glazing compound, 78; *ill. 80*
Government standards, regulations, and programs, 156–70

Heat exchanger:
  air-to-air, 22
  for hot water tanks, 139; *ill. 140*
Heat flow, 16–56
  basic equation, 29
  calculating, 28–32
Heat loss:
  calculating, 28–32
  chimney, 20, 21–2
  cost, 30–2
  detection, 10–14
    infrared scan, 11–12; *ill. 13*
  doors, 21
  hot water tank, 140–3; *ill. 141, 143*
  house trim, 63–4; *ill. 63*
  siding, 64–5
  vents, 20, 21–2
  windows, 2
Heat transmission, 5–7
High-R Shade, 41; *ill. 41*
"Hole-plugging," 57–83
Home Energy Savers Program, 160
Hot air ducts:
  insulation, 151–2
    amounts recommended, 152; **table 152**
Hot water tank:
  heat loss, 140–3; *ill. 141,* **table 143**
  insulation, 138–49
    cost factors, 143–6; *ill. 145*
    electric heater, 146–7; *ill. 148*
    gas heater, 147; *ill. 148*
    Sepco water heater, *ill. 150*

Index

manufacturers' improvements, 148–9
House trim:
   heat loss, 63–4; *ill. 63*
HUD, *see* Dept. of Housing and Urban Development
"Huff and puff" (infiltration), 1–2, 10, 20
   see also Heat loss
Humidity in airtight house, 49–51

Infiltration ("huff and puff"), 1–2, 10, 20
   see also Heat loss
Infrared scan for heat loss, 11–12; *ill. 13*
Insulation:
   blower, 110; *ill. 110*
   definition, 3, 7–8
   fire safety, 51–2
   hazards, 14–15
   materials, 84–98
      cutting, 99–102
      effect of precipitation on, 48–9
      effect of water vapor on, 46–8;
         see also Vapor barriers
      flammability, 51–2; cellulose, 89; fiberglass, 87–8; perlite, 51, 97–8; polyisocyanurate foam, 92; urea formaldehyde foam, 51, 95; vermiculite, 51, 97–8
   labeling, 165–6
   manufacturing processes, 9
   prices, 91
   protective measures for users, 99–100
   R-values, 23–32
   for water heaters, 147–8
   see also specific materials, *e.g.* Fiberglass
   in new buildings, 128–37
      attic, 136–7
      basement, 128–30; *ill. 129, 130*
      cathedral ceiling, 137
      floors, 131–2; *ill. 131*
      foundations, 128–30; *ill. 129, 130*
      walls, 132–4; *ill. 133*
   quantity required, 32–5; *ill. 33*
      for air ducts, 152; **table 152**
      for basement, 35–6
      for ceiling, 35
      for walls, 32–5; east wall, 37; south wall, 36–7; west wall, 37
   retrofitting, 99–127
      in attic, 50, 102–5
      crawlspace, 119–20; *ill. 120, 121*
      floors, 117–18; *ill. 118*
      roof, 115; *ill. 114, 116*
      walls, 116–17
   state policies and legislation, 168–70
   tax credits for, 163–4

Knee walls, 112–15; *ill. 113, 114*

Labeling of insulation materials, 165–6
Loose-fill, 89, 102
   blowing in attics, 110–12; *ill. 110*
   hand-pouring in attics, 108–10; *ill. 109*
   vapor-barrier for, 108; *ill. 109*
   see also Perlite; Vermiculite

Masonry pointing, 67–70; *ill. 70*
Mineral wool, 88
Model Energy Conservation Code, 166–7
Mortar:
   for masonry pointing, 68–70

National Association of Home Builders, 50
National Conference of States on Building Codes and Standards (NCSBCS), 166
National Energy Conservation Policy Act (1978), 158, 168
   programs, 158–61
National Fire Protection Association, 44, 95
New buildings:
   codes and standards, 156–8, 166–8
   insulation, 128–37
      attic, 136–7
      basement, 128–30; *ill. 129, 130*
      cathedral ceiling, 137
      floors, 131–2; *ill. 131*
      foundations, 128–30; *ill. 129, 130*

walls, 132–4; *ill. 133*
Newspaper:
  as filler before caulking holes, 60

Oakum:
  to caulk foundation sills, 63
  as filler before caulking holes, 60
Odors in airtight house, 22–3

Perlite:
  as attic insulation, 109–10
  flammability, 51, 97–8
  manufacturing process, 9, 97
Permeability of building materials, **table 48**
Perm rating, 47
Pipes:
  insulation, 149–51
  see also Air ducts: insulation
Plastic film, *see* Polyethylene film
Pointing masonry, 67–70; *ill. 70*
Polyethylene film as vapor barrier:
  for crawlspace, 119; *ill. 120*
  for floors, 117
  for loose-fill in attics, 108; *ill. 109*
  for new foundations, 129; *ill. 129*
  for new walls, 134
Polyisocyanurate foam:
  flammability, 92
  manufacturing process, 9, 92
Polystyrene foam, 9, 89–91
  extruded form:
    for below-ground use, 36, 129
    flammability, 90
    manufacturing process, 9, 90–1
  flammability, 51
Polyurethane foam, 8, 91–3
  effect of sunlight on, 92
  flammability, 51, 92
  R-value, 91–2
Post and beam house:
  insulation, 134–5; *ill. 135*
Precipitation:
  effect on insulation, 48–9

Radiation, 6–7
Reflective insulation:
  between radiator and wall, 96–7
  to increase R-value of air space, 25

Refrigerator:
  insulation, 153–4; **table 153**
Residential Conservation Service, 159–60
Retrofitting insulation:
  in attic, 50, 102–15
  cellar, 119, 121–4; *ill. 121, 122, 123*
  crawlspace, 119–20; *ill. 120, 121*
  floors, 117–18; *ill. 118*
  roof, 115; *ill. 114, 116*
  walls, 116–17
Rock wool, 9, 88
Roof:
  framing, *ill. 100*
  insulation, 115; *ill. 114, 116*
  see also Cathedral ceiling
Rural housing weatherization program, 160–1
R-value:
  affected by drainage, 36
  calculating, 28–32
  principle, 8
  of various insulating materials, 23–32; **table 25–7**
  of various wall constructions, 52–6
  of windows, 38–40

Safety measures for use of insulation materials, 99–100
Scan for heat loss, 11–12; *ill. 13*
Sepco water heater, *ill. 150*
Shurcliff, William, 43, 44
Shutters (thermal), 2, 41–3, 44–6; *ill. 42–3*
  homemade, 43–4; *ill. 43*
  mounting, *ill. 45*
Siding:
  heat loss, 64–5
Sills (door):
  leveling, 74
Sills (foundation):
  caulking, 62–3; *ill. 62*
  sealer, 132; *ill. 132*
Slab (concrete):
  perimeter insulation, 130; *ill. 130*
Slag wool, 88
Smoke detector, 45–6
Solar energy equipment:
  tax credits, 164–5

Index

South wall:
  insulation, 36–7
South windows:
  shutter management, 46
  as solar collectors, 36–7, 40
State conservation plans, 166–70
State Energy Conservation Program, 157–8
Storm doors:
  caulking, 65–7
  tax credits for, 164
Storm windows, 24
  caulking, 65–7; *ill.* 65, 67
  tax credits for, 164
Styrofoam, *see* Polystyrene foam: extruded form
Sunlight:
  effect on foam, 92, 129
Sweep, 76; *ill.* 75

Tax credits for:
  insulation, 163–4
  solar energy equipment, 164–5
Temperature:
  balance temperature, 16, 17–19
  cooling balance temperature, 18–19
  definition, 5
  dew point, 46
Thermal bridge, 31–2; *ill.* 31
Thermal curtains, 41, 43, 44–6; *ill.* 41
Thermal shell of house, 19–23
Thermal shutters, 2, 41–3, 44–6;
    ill. 42, 43
  homemade, 43–4; *ill.* 43
  mounting, *ill.* 45
Threshold gaskets, 73–4, 76; *ill.* 75

Urea formaldehyde foam:
  danger of, 93, 98n
  flammability, 51, 95
  manufacturing process, 9, 93
  shrinkage problem, 93, 94–5;
    ill. 94, 95
Urethane foam, *see* Polyurethane foam
Utilities (gas and electric):
  energy-saving obligations, 159–60

Vapor barriers, 47–8
  cause excessive humidity, 49–51
  for cellulose in walls, 117
  for floors, 117
  for loose-fill in attic, 108; *ill.* 109
  for new walls, 134
  see also Polyethylene film as vapor barrier; Water barriers
Ventilation:
  attic, 50
Vents:
  heat loss, 20, 21–2
Vermiculite:
  as attic insulation, 109–10
  flammability, 51, 97–8
  manufacturing process, 9, 97
Vestibule to protect door, 21, 22

Walls:
  basement:
    insulation, 121–4
    framing, *ill.* 100
  insulation:
    in new buildings, 132–4; *ill.* 133
    quantity required, 32–5
    retrofit, 116–17
    knee walls in attic, 112–15; *ill.* 113, 114
  north-facing:
    caulking holes, 66
  R-values of various constructions, 52–6
  siding:
    heat loss, 64–5
  south-facing:
    insulation, 36–7
Water barriers:
  in basement construction, 128–9
Water heating, *see* Hot water tank
Water pipes:
  insulation, 149–51
Water vapor:
  effect on insulation, 46–8, 50–1
  see also Vapor barriers
Weatherization program, 158–9
  rural, 160–1
Weatherstripping, 20, 21
  doors, 70–7
    materials, 70–4; *ill.* 71, 73; kit, 73; *ill.* 75; shoe, 74, 76; *ill.* 75; sweep, 76; *ill.* 75

methods, 74–7
materials, 70–4; *ill. 71, 73*
tax credits for, 164
windows, 79–80, 83; *ill. 81, 82*
Wind:
  effect on heat loss, 7
Windows:
  basement, 130
  casement:
    weatherstripping, 79–80
  caulking, 58–61
  double-hung sash:
    weatherstripping, 80; *ill. 81, 82*
  double-pane, 24; *ill. 24*
  heat loss, 2
  calculating, 28, 29, 30, 38–9; *ill. 38*
  heat transmission, 37–46
  prefabricated, *ill. 59*
  reglazing, 78–9
  R-values, 38–9
  as solar collectors, 36–7, 40
  south-facing, 36–7, 40
    shutter management, 46
  storm, 24
    caulking, 65–7; *ill. 65, 67*
  see also Thermal curtains; Thermal shutters
Wiring, *see* Electrical wiring